(ART)iFACT

RE-RECOGNIZING THE ESSENTIAL OF PRODUCTS

viction:ary

(ART)IFACT

First published and distributed by
viction:workshop limited
URL: www.victionary.com Email: we@victionary.com

viction:ary™

Concept and art direction by Victor Cheung
Book design and produced by viction:design workshop
Edited by viction:ary

ISBN 978-988-98228-6-6
Printed and bound in China

Key to symbols:

kitchenware
furniture
clothing + accessories
out-door accessories
personal accessories
lighting
home decorations
household commodities
home electronics

CONTENTS

PREFACE

Page. 004-008

OBSERVATIVE + AUGMENTATION

Page. 009-088

FUN + FUNCTION

Page. 089-168

INTERACTIVE + SENSATION

Page.169-248

BIOGRAPHY

Page. 249-262

PREFACE

[Art]ifact has sought to bring appropriate works into pages with a particular character. A palette of designers generating the intellectual impetus to carry their study of the meaning of objects into the territory where design meets art. An open space for reflection, by grouping Droog's personal and inquisitive objects with Martin Guixé works on living matter and Pascal Anson's inspirations from unconscious design to create design objects that are "different, challenging and beautiful." "A kind of miser", Anson turns ordinary materials into magical objects. Connected with one another in climatic zones, as zones of influence in contemporary design, stories are told while objects leave their own questions.

The significance of [Art]ifact lies not so much in the body of work itself, however much you admire the creativity of the individual pieces, but in the resonant ideas that underpin them. One of the fundamental notions in [Art]ifact is the idea that design itself can be a critical medium for cultural reflection without recourse to interpretation by the design critic who writes about the artefact.

Despite the weight of evidence that art and craft sit along the same spectrum of creative visual art, the divide persists both bureaucratically and socially. Its persistence has more to do with retaining the status accorded certain possessions and ways of working than it does with a genuine difference between the fine artist and craftsperson. [Art]ifact begins at this point. It is not interested in determining where the borders might lie but in showing work which ignores all categories as unhelpful to looking at things. It is concerned to get out with the job of being stimulating, poetic and though-provoking. In doing so, the wish is to contribute to the creation of a forward-looking art and a more open minded approach to all forms of visual practice.

'Art is not a profession. There is no essential difference between the artist and the craftsmen[1]', says Walter Gropius. Borders are never comfortable places to be, but they are often where the most interesting things happen. Nowhere are the boundaries more volatile and complex than those that lie between art and craft and between design and craft.

The strength of the Bauhaus was that it turned its eyes forwards. In order to prevent the artist being shut off from the world of industry and handicraft (and vice-versa), it was the first art and design school to make a point of teaching the specific disciplines of metalwork, carpentry, weaving and ceramics alongside a rigourous programme of art theory and exercises in manipulating colour, line and form. The results of this aesthetic cross-fertilisation included the influence of Paul Klee in the stained glass and weaving workshop, the blueprints for adjustable metal lamps and coffee pots from painter Marianne Braundt and the highly regarded typefaces of photographer Moholy Nagy.

The Bauhaus designed objects which are still admired and produced today, form a potent example of how and why art-craft demarcations could and can be mutable and how this hybrid might be adapted to face the future.

It's significant that craft and art also stood shoulder to shoulder at another crucial moment in the evolution of 20th Century Modernism. The Constructivist artists who provided the virtual accompaniment to the 1917 Russian Revolution sought to destroy hierarchies in society. 'Go into the factories, this is the only task for artists[2]'.

For everyone of the designers in [Art]ifacts the border between art and craft has become their homeland. In their richly varied work there is no sense of transgression, but rather a transcendence of restricting categories into art forms that stand in their own right and require no labels. In each case, a background in individual craft practice is just one of many elements and influences which each combine to make the finished work a demonstrative of both skill and sensitivity to the environment and understanding of current art theory and practice.

Design is alive, well, morphing, evolving, becoming more complex, more pervasive, and evident in the lives not only of the experts but in the lives of people. Design is no longer some distant, half-imagined, half-understood poor relation in the business process. Contemporary design, a timeless concept, exemplified best by an object that is soundly manufactured and beautiful, works efficiently for its purposes, and suggests ideas that transcend its form and function. Design is about 'products worth bonding with for a lifetime,' says Wanders. Ours times are of concern and awareness. Stimulated by a spreading preoccupation with the world that surrounds everybody, contemporary design should be valued for "lightness, speed, accuracy, visibility, multiplicity and consistency[3]" as Italo Calvino suggested for this Millennium.

Finding these qualities is a shared responsibility: all those with power need to use that power wisely and responsibly to this end, where they can. Designers, through their designs, exert a powerful influence on society. If we take this ethical responsibility seriously, we will want to be active as protagonists in shaping the future, making sure that the results of our work advance people's personal growth, create a healthy living environment, both natural and artificial, and engender a relation between people and their environment that is maximally harmonious.

Objects remain important as a symbolic location of experience but a shift is occurring in Contemporary Design. The attention is focused on the psychological, social and cultural contexts that give meaning and value to products and to the discipline of design practice. This shift is guided by the recognition that objects are situated in a variety of ways that were not clearly understood or adequately explored by the designers, historians, critics, and theorists who established the precedents for the field of design earlier in the 20th Century. As a consequence, contemporary design discourse does not represent a radical break with the past. Instead, it represents a broadening of the scope of design to include issues and problems that were often embedded in earlier practice and speculation but were seldom developed in depth.

The most interesting as well as the most difficult thing about contemporary design is that you can never really put your finger on it. We may understand or identify it but usually "only" emotionally. Contemporary design can be baffling and lyrical, provocative and disturbing.

Beauty, thinking, social, excessiveness, allegory, aesthetics, poetry are a few words that can describe the works presented in [Art]ifact. Art or Design? Art or Craft? Design or Craft? Above all, they are pieces of communication, be it physical, psychological or both. Now the frontiers are being breached. This time, practitioners working in all media are challenging distinctions not by issuing angry manifestos and storming the barricades, but by moving comfortably and constructively between different camps.

This book is not an effort to establish a monistic vision of design. It is not based on a single unifying concept of design or on the prospect of a single philosophy of design. Rather, it is based on a belief in the value of radical systematic pluralism: the principled cultivation of a sustained conversation among individuals with widely differing perspectives on the natural and human-made world. The element that unites the contemporary designers work presented most of all is the pluralism. A pluralism that keeps alive the ongoing search for truth and understanding by focusing inquiry on common problems encountered in experience. In addition to a search for identity within a context of globalization comes into their work. Not all of these designers are formally associated with one another. Some belong to separate generations or have attended different schools and come from different parts of the world. Nonetheless, they all share a similar sensitivity. The value of those objects is beyond money, form and function. In the face of a throwaway culture that consumes meaningless products, contemporary designers want their creations to have more quality and more personality.

[Art]ifact challenges not only the dictum of 'An object intentionally made for a certain purpose[4]' but more contemporary notions of design such as problem solving or commercial branding by refracting Modernist design principles through the artist's eye.

Idealism without practical application is worthless. The system within which we live requires of industrial designers that they ensure that their products are commercially successful. The companies for whom they work exist to provide investors with a good return on their investment. An ailing company is unable to help anyone in society: it is a burden to all. design products should help people accomplish and experience the things they want to accomplish and experience; they should allow them to do these things with minimum effort and in natural, intuitive ways; and, in their manufacture, operation and effects, such products should be fully in live with their users' fundamental value systems.

The conjecture is the creative part and designers are very good at that. But then researchers must do their best to refute their own ideas. This is where many designers fall down, lacking the in-built scepticism to challenge their own speculations. This palette of designers does not suffer from this syndrome. Taking experimentation as design research method, ideas are subjected to a critical process of refutation, always questioning the work and using scholarly approach to the milestones of Modernism to enhance its depth.

One could describe this approach as somewhat introspective but it is also full of insight and engagement. [Art]ifact captures exceptional practitioner/researchers in their prime. These products and interiors mark a cross-cultural shift that goes beyond bricolage to open new design possibilities. There are forms that go beyond functions, beyond aesthetics. Understand that sandpaper is a must-use item among the product design industry; the book cover is chosen to be made of it. Not only because of its familiarity that could link the title more closely to the industry, but also the nature of it that is interesting in explaining the attitude more concretely. And now, please feel free to see beyond.

Luísa Santos

Designer and MA
in Curating Contemporary Art,
Royal College of Art, London

Notes:

[1] Programme of the Staatiche Bauhaus in Weuimar, 1919 reprinted in Haus M Wingler, The Bauhaus, MIT, 1980

[2] Georgio Vasari, 'Lives of the Artists', Penguin Classics, London, 1981

[3] Ítalo Calvino, 'Seis Propostas para o Próximo Milénio', Relógio D´Água, Lisboa, 1998

[4] http://www.wikipedia.org

OBSERVATIVE + AUGMENTATION

CONTEMPORARY DESIGN, A TIMELESS CONCEPT, SUGGESTS IDEAS THAT TRANSCEND ITS FORM AND FUNCTION. IT IS IN MANY WAYS A SELF-CRITIQUE OF CONTEMPORARY DESIGN THAT ATTEMPTS TO GIVE A PASSIVE RESPONSE TO MASS CULTURE. AS IN MUSIC, MAKING A MODIFICATION OF A THEME BY INCREASING THE TIME VALUE OF ALL ITS NOTES, THE AUGMENTATION IN THE FOLLOWING DESIGN OBJECTS REVEAL DETAILS THAT ARE OFTEN OVERLOOKED BY THE COMMUNAL. THESE OBJECTS ARE A PROVOCATION TO HUMAN SENSES, ASKING US TO SEE BEYOND THEIR [APPARENTLY] SIMPLE FORMS AND FUNCTIONS.

009 I.V. DRIP FEED PLANT POT

This self-watering plant pot's main feature is a medical drip feed. This unique characteristic allows the plant to be nourished with water and food as it is required. The drip feed provides a striking design focus as well as acting as a reminder to its owner as to when it needs topping up.

010 NIKLAS

Niklas is a porcelain sphere which is glazed on its inside. It has a spacious opening with a mushroom like sector to hang it up and a plane foot.

The sphere can be used as a lampshade just by hanging it up over a common bulb and its cable.

All kind of foods and dishes are presented and staged beautifully due to the uprising backside of Niklas. If salad servers or soup ladles need to be used, they can be arranged practically inside the mushroom like sector.

Niklas also serves as a flowerpot, standing on the gound or being hung up with a knot in the hanging rope.

Apart from many other ways of using Niklas (arranging buffets, birdhouse for the garden) one is still worth mentioning: Its use as a lantern, standing or hanging inside the house or in the garden, preferably in a little crowd, with each of them glowing softly in the dark.

012 CANDLESTICKMAKER

The CandlestickMaker, a 'relative' of the VaseMaker, is a processed fragment. Born from a dissection of a basic candlestick structure, it is a fragment that has left behind its supporting leg. It reinvents itself each time it's combined with a new supporting object which functions as the missing limb.

Descriptions: Page:

VASEMAKER 013

VaseMaker dissects the classic vase into its two functional elements: Support for the flower and a container for the water that sustains it.
VaseMaker is designed and redesigned each time it is placed upon a vessel. A symbiotic relationship is created between the iconic shape of the VaseMaker and the form of any ready-made vessel that is placed underneath it. VaseMaker is a half-function object that leaves space for the imagination and for the participation of the user.

WATERPROOF 014

Based on the model of the Wellington boot, the designer aimed to retrieve this cheap, cult, maximally functional footwear from oblivion and to immortalize it in the form of a porcelain object, a vase. He tried to invert Sullivan's cliché of form follows function into function follows form. He inverted its defining role as an object protecting against water into an object containing water. He also changed the material from the cheapness of rubber to the costliness of porcelain. And then, a boot becomes a vase. One shape ends up in a different context and acquires a new function. When this object is multiplied, various compositions evoking diverse associations are possible.

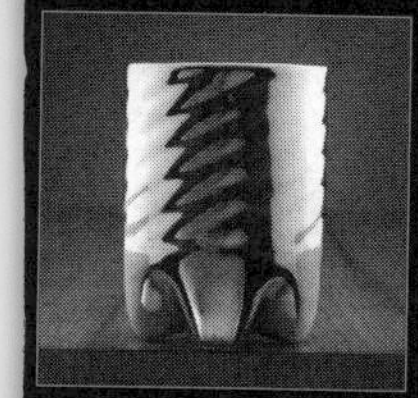

COLA CUP, GOOD WATER CUP (GOLD, WHITE, BLUE ONION EDITION) 015

Inspired by fast-food culture, the changing lifestyle on the path from communism to capitalism, Maxim Velcovsky began to cast sundry recyclable shapes. Amazed by their minimalism and fragile aesthetic, so close to porcelain, he began to create a collection for fast people who kept their services locked up in cupboards at home and spent their days in offices, consuming food from instant, single-use design. He was fascinated by the fact that, by a mere change of material, one could draw attention to a shape that people discarded every day. Inspired by a natural passion for DIY (the cut-off bottoms of PET bottles are a handy universal substitute for any kind of vessel), he cast a series of recyclable design and suspended it in time in the form of fossils, in the noblest material of all, porcelain.

Page: Descriptions:

020 POWER-CUT SOLUTIONS

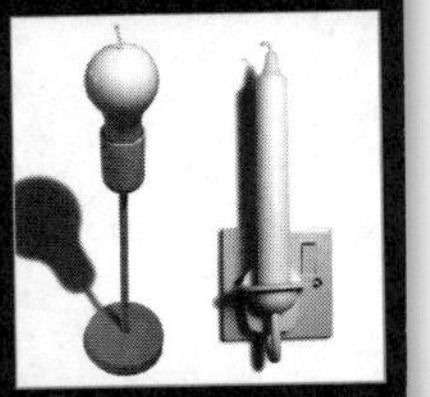

Power-cut solutions includes E27 candle and Power-cut Plug – two bright ideas to keep living space illuminated, even in a black out.

021 ICON CANDLES

Icon Candles includes Coke candle, Water candle and May Day candle. Madam Tussaud would love them. Let's light up the wax casting of well-known supermarket icons and make a time for meditation.

022 STAK

Designed by Karim Rashid for Danese in 2003. Stak is the perfect stackable modern table set. Each set includes a plate, bowl, saucer and cup. Made of high quality Italian ceramic-ware.

Colour Options: White, Black, Red

023 MORPHSCAPE

'The non-stop continuation of a single undulating surface, divided into every need for the table as a modular scape. The inspiration is the city scape of Istanbul from minarets to mosques to the modern city and the Bosphorus,' Karim says.

It is made in Germany, with white glazed porcelain. There are different sets of Morphescape carrying special names from Istanbul: 'Kumkapi' tapas set; 'Tunel' service set; 'Bebek' tea/coffee set; 'Kavak' dinner set; 'Bazaar' snack set; 'Galata' wine & cheese set and 'Halic' decorative set.

Descriptions: Page:

DISH NO.3
024

Dish no. 3 is an exact copy of a tray that resembles a somewhat excavated slice of a ball and that had been produced earlier by Casimir in solid oak. But while the other materials were surprisingly heavy, this one is remarkably light due to its material - EPP - Expanded Polypropylene.

BASKET BOWL
026

A casting of an official sized deflated Basketball, intended to function as a fruit bowl. Limited edition of 50 produced by designer Alex Garnett.

FOOTBOWL
027

A casting of an official sized deflated Football, intended to function as a fruit bowl. Limited edition of 50 produced by designer Alex Garnett.

JAMES
028

James is a plate for events like openings, vernissages and stand-up parties, which means for occasions, where drinks and appetizers are served with style.

A handy plate, that takes the food in its glazed hygienic inner area and holds all glasses with a handle within its special hollow. Due to its broad and unglazed edge, it provides good grip. Everything in one hand, guests always have a hand left to eat or to say `hello´ to their friends.

Page: Descriptions:

030 SOMMELIER

Sommelier is a very hand-made vine glass inspired by disposable fast food shapes.

031 OH! MY GLASS

Oh! My Glass - The phrase 'from top to bottom' has never been so playfully enacted. Oh My Glass gives every reason for another 'fill'er upper', a perfect way to say 'bottoms up' in sheer style.

The nifty glass that also serves as a bottle is every thirsty person's best friend. With a quick flip of the glass and a flick of the wrist, you'll be enjoying that cold beer before you can say 'oh my glass' and not to mention, before everyone else.

Oh! My Glass will leave you screaming in delight, as there's never been a trendier way of enjoying your favorite drink. Share a delectable bottled beverage with that special someone tonight... all you'll need is some candlelight and that special something to make their heads spin with delight.

032 UNDINE

Water, an amorphous nomad, at some point in time between mountain and stomach in the Undine, forced into oriental surroundings. The Undine is filled from the top; water is poured from the neck. Upturned glasses close the receptacle. It can be carried in one hand.

TWIN 034

Twin makes it possible for two people to enjoy hot or cold meals / drinks. It is possible as well, to enjoy two different ingredients at different times. It is a thermos receptacle consisting of two half bowls which are connected with the help of just two magnets which are embeded on the inside of the rubber coating of the lids.

Because of carbon dust included in the vacuum and the bowl like shape - hereby the ratio of surface to volume is minimal - the temperature is being preserved much longer as with standard insulating techniques and a high protection of damage is provided.

The magnetic connection of the bowls, as well as the 'clipclap' function of the lids, which have to be pushed onto the green spot in their middle to open and at their sides to close, a very easy and precise use is possible, even with cold hands, with gloves or with just one hand.

Additional to that, both mechanisms provide a better hygiene, because of their simple form without threads. The big openings of the bowls are easy to clean and even the lids are dishwasherproof.

At the same time the lids can be used as saucers, as well as the grooves on their upside provide space to take up the strapband with its adapter. The used materials are stainless steel (bowls), steel and silicon (lids).

GLASS TOASTER 036

It is a toaster that made out of glass, which celebrates toasting in a glowing and shining way that makes us look forward to enjoy a fresh piece of toast.

It explains how someone works. He neither hides the bread, nor its preparation. In this way it is possible to receive him as an idyllic little light. He stages the bread and the way it gets toasted, which is accompanied by warm light and a tempting aroma. Hollows at both sides allow taking hold of the bread without touching the toaster itself. The switch is designed like a remote control that also houses all electronic control components.

BUON APPETITO 038

The bed linen 'Buon Appetito' is inviting you to do things that you normally shouldn't do: eating in bed. At the same time it is also functional, because the crumbs will not fall on the mattress. Reminding you at a typical Italian restaurant the big red and white bib just makes you smile.

(Referring to the picture with a lady) While Lucy is usually having breakfast in bed her appetising bedcovers protect her from the little mess making monsters spreading all their crumbs.

Page: Descriptions:

040 PASTA FORK

This design for a pasta fork resulted from a competition brief set by pasta company Barilla. They were looking for innovative new designs for forks for one particular kind of pasta. After some experiments the trident form was chosen to solve the problem, especially apparent with penne, of the pasta slipping from the fork once pierced. By adding spear-shaped tines, the elasticity of the pasta helps keep it on the fork. The lines of the fork make no attempt at a re-styling of the archetypal form, trying instead to embody it.

041 MP SPOON

Series of two silver sugar spoons, inspired by the classic Dutch spoon 'Hollands Glad'. One of the spoons is titled MP, and was designed by Ed Annink for the Dutch Prime Minister. It was commissioned on the occasion of the Dutch EU presidency in 2004 and was offered as a promotional gift to foreign government leaders. Whereas in the past sugar spoons were used to bring sugar from the sugar bowl to the cup, sugar cubes and sugar bags have made this use redundant. A hole has been made in the head of the spoon, as to emphasize the spoon is solely used for stirring.

041 DB SPOON

Series of two silver sugar spoons, inspired by the classic Dutch spoon 'Hollands Glad'. One of the spoons is titled MP, and was designed by Ed Annink for the Dutch Minister President. A second variation is titled DB. Starting from the same base form, an open decoration has been made, inspired on Delft Blue decoration.

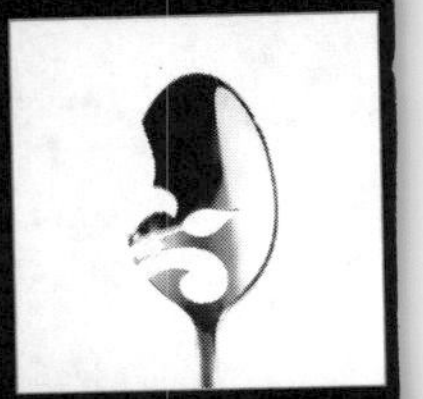

042 BRUSH COUPLE

It's an inseparable brush. Connected with magnet, they cannot be hanging alone.

Materials: Wood, Horse hair, Aluminium & magnet

Descriptions: Page:

MATCHES 043

Variation around matches and the habits they create.
3 Life: Allow three chances to get the fire on / Construction: Create big cathedral with… / Toothpick: after dinner / Candle: a minimum chandelier / Coffee: matches for after dinner cigarette habit / Muscle: a strong one / Incense: for bathroom / Wind: solution to locate where the wind come from / Single: A precious one made out walnut wood with it's own container.

WOOFER 044

This is functional kitsch; the wrong becomes the new right. By adding a function to an otherwise grotesque object, it acquires new aesthetic values, becoming an object of desire. Pun intended, this 'Woofer' speaker system holds is a cross-over between an addition to your sound system and your loyal 4 footed companion. Available in a co-axial speaker system version (two dogs), and soon followed by our tweeter speaker system for your desktop. Of course this speakers tem, with it's unusual set-up, is not to hold up against a high-end/high costs speaker system, but engineered by a professional audio studio it sounds remarkably well, compared to an everyday speaker system!

ROLLING 045

It's a research around the ceramic extrusion. The result is an amazing series of container with cork caps.

This work has been done for the master diploma of industrial designer at Ecal Lausanne.

Materials: White glazed ceramics with cork

OH CHINA 046

It is a response to the prevalence of mass-produced product, its ultimately disposable nature and the sense of the loss of the individual maker. The spool of thread is a symbolic mass-produced product, however each one is totally different to the other, either in colour, weight, of the way the interior spine becomes squashed. The fact that it is then taken and further applied to another product in the form of clothing, perhaps again factory produced, further reinforces the contradictory notion of individuality in an increasingly homogenized product driven world. The realm of the mass-produced is now extended to the individual maker / craftsman.

Page: Descriptions:

048 GHOST TREE

The designer wanted to incorporate the nature into his design work as a reaction to sleek forms created on computers. He wanted to infuse his pieces with life and humor, but still tie into a minimalist aesthetic.

The ghost series was created by appropriating imagery from the outdoors and sixties modernism. The originals were created using a process similar to collage. Started with a tree branch, caribou hooves, or a set of deer antlers, the designer then sculpted the pieces around them.

050 ROEDEER, DEER, MOOSE

French by origin, Augustin Scott de Martinville lives in Lausanne, Switzerland where he studied at the famous local design school, Ecal. His aim is to edit each new year a series of objects that are linked to a self-imposed theme. The first year's theme is the reinterpretation of the codes of bourgeois furniture. The Moose & Deer Head are part of this series: a modernised, somehow abstract version of the trophy head, executed in triplicate. 'We wanted something light to stress the heavy drama that is usually associated with this object – the killing of the animal and the stuffing of the head.'

052 HEATWAVE

In a time of arbitrariness and opportunism, the designer is searching for ornamental but somehow necessary form. Modernistic functionality and post modernistic effusiveness do not have to exclude each other. He wanted to demonstrate that functionalists are also sinners of styling, and that soberness is not always more functional than highly decorative form. A radiator needs a surface as large as possible to lose its heat radiation. Ornamental and highly decorative form has a surface that is larger than sober conventional radiators. Using this aspect, he designed a decorative radiator that is functioning better or as good as a normal heating.

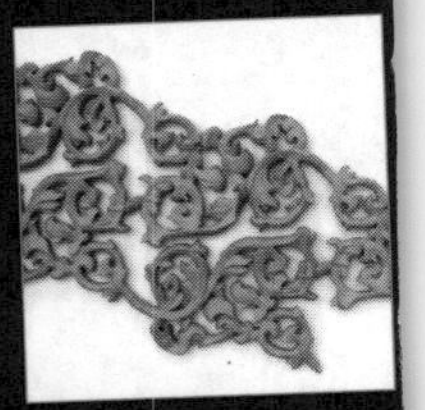

054 ROADKILL CARPET

The Roadkill carpet is a continuous struggle between attraction and revoltion. It's a warm, soft, cuddly carpet that attracts you to take a nap on it. On the other hand, it's an image of a flat, bloody fox. Something you do not want to be nearby.

The designer is always fascinated by things which people actually do not want to see. Do you know the feeling that you do not want to take a look at something, but you still do? (Handmade carpet from 100% wool)

SPRUCE – SWEEPER LIGHT 056

This free-standing lamp doubles up as a floor sweeper. The sweeper is integrated into the base of the lamp, allowing consumers to give their homes or a particular dirty spot a quick once-over without the hassle of having to engage in a full cleaning session. The lampshade is flocked with a dustmite pattern which can be seen once the light is switched on.

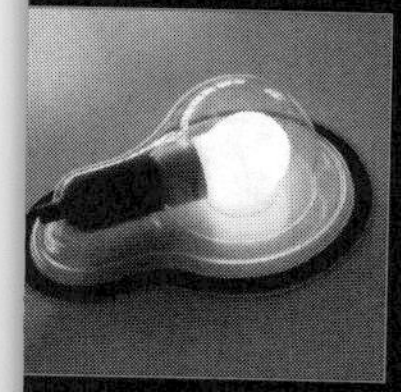

STICKY LAMP 058

Sticky lamp is a lamp like a sticker. Peel off the back lid and stick as high or low as you like. It's the outcome of a research into the aesthetics of instant solutions; the things people tend to resort to when something breaks down, or lacks something, and needs to be resolved instantly. The designer found it very intriguing of what tools were used to repair products: tape, rope, string, rubber bands, paper clips. One is about a lamp fitting that was taped against the wall to provide instant light. He then translated this into the Sticky Lamp. It still has the immediacy and randomness of the original, but has been transposed into a sellable product.

SNOTTEBEL 060

Snottebel is a series of decorative and ambient lights made from LED light chains mounted on an aluminium rod dipped in translucent white rubber.

The rubber forms an embodying skin around the whole structure. The LEDs shine through the white rubber. All lights are dipped as a straight line to enable mass production. Once the rubber is cured, the lights are distorted in their unique branch-like shape. Snottebel is designed to be hung from a ceiling or a wall.

1TOTREE 062

A lamp made up of similar elements. It is extendable from simple lighting to enormous chandelier by adding more elements.

Page: Descriptions:

064 SUPERORDINATE ANTLER LAMPS

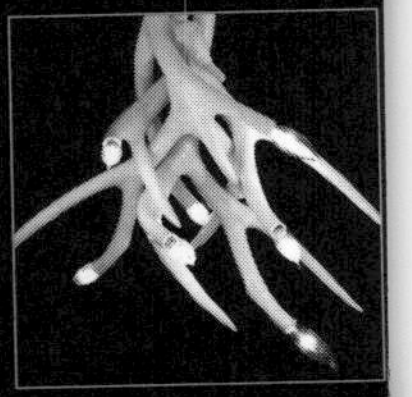

We can rebuild it better, stronger, faster. What we find is not good enough. We make things because we can make them better. And thus the Superordinate Antler Lamps are nature, only better. Superordinate Antler Lamps are lamps made from deer antlers cast in ceramic.

Materials: Ceramic

Illumination: Multiple 25 Watt Incandescent Bulbs

066 SPAGHETTI CHANDELIER

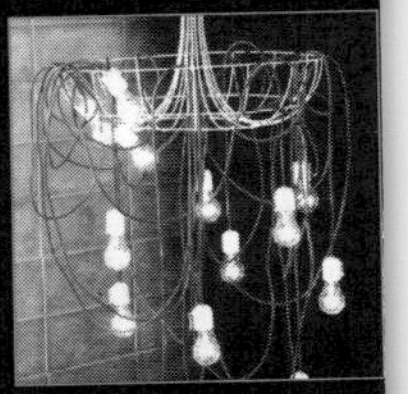

Spaghetti Chandelier is a random decoration pendant light. It is decorated with red-coloured electric wire on white powder coated metal wire frame. The highlight of it is the 'orange chord', which is widely used as an industrial material. He tried to transform the industrial material into the design of domestic furnitures and products. 'The special socket' is another design concept he tried to introduce through this project. It has a hook that enables the lighting to be hung to the wire structures in anyway the consumer may please with.

Materials: Powder coated steel, Electric wires.

067 ELECTRIC SPAGHETTI PRINCESS CHANDELIER

Spaghetti Princess Chandelier is made of Perspex, electric flex and light fittings. It is available in Black or White electric flex.

'Our inspiration came from the beautiful chandeliers of the 17th century Baroque era, and we have tried to capture an essence of their silhouette and feeling of continues movement using the electric flex itself. Its a subtle reminder of the lavish exuberance of times past designed to fit in contemporary interiors.' Brendan Young.

068 WORLD VIEW

World View is a virtually weightless lighting fixture. Constructed from a cluster of 91 LEDs suspended from a silver helium balloon and anchored by a spool at floor level, World View proves the simplest, most humble materials can reach great heights.

Designed with public spaces, special events and fairs / exhibitions in mind, World View is currently being developed for mass production. A limited edition version can be viewed at the Gallery space of sketch, London from October 2003.

Descriptions: Page:

POODLE

070

The Poodle is chairs that made with interior textiles. The main chair is a raw, unshaved model, which can be shaved in any model you wish. Just like the shaving and modelling of a real Poodle, you can design this chair on your own.

BATH & BEYOND

072

Bath & Beyond is a readymade bath chair. Cut, bent and legs added - the bath becomes a surprisingly comfortable chair. It is great for indoor use and perfect for sunbathing outside. Working on this readymade object, made the designer think of the connection between the initial use of the object and the outcome. Their intention was to make the sitting on the Bath & Beyond fill oddly familiar. From the smile on people faces they think it was well accomplished.

Materials: A bath, Steel

BROOMSTOOL

074

A four-legged stool, which cannot deny its origin, emerged from the surprising and unusual combination of four brooms. Yet undiscovered material properties are being emphasized and made use of. The innovative use of bristles as upholstery creates an absolutely new and stunningly comfortable comfortable sitting sensation. The broom sticks form a classical fundamental structure in which the conventional screw fixing between broomstick and bristles constitute the feet of the stool.

Old name : Besenhocker

BATHROOM

075

A seating group evinces a puzzling character that invites the viewer to experience it sensually and reflect on its origins. The forms are based on a Plexiglas bathtub and shower tray ensemble. The bathtub has been divided into three objects, with nothing lost in the process. All four pieces are set on stainless steel legs. The focus is on the alienation of bathroom furnishings from their customary surroundings and function.

Old name: Die offene Gestaltung

Page: Descriptions:

076 FALB

Falb is a chair which offers some space to place one's personal things. One stays near it and still has enough room. Its shape differs from the symmetry of usual chairs. Its back carries jackets, borses, ecc. In order to take weight its right chair-leg makes a side-step. It shows its individual character.

078 BIG SOCKS, PICK POCKET, COSY COVER

dressing - Clothing for Chairs

Clothing adds something to the body without reducing its original quality. dressing transfers this principle to furniture in order to add a new quality. A collection of clothes for chairs has come into existence that supports their usage and even extends it. dressing can be put on and taken off just as you like. When not used, these clothes can be folded up and stored in a wardrobe just like ordinary clothes. The collection will consist of four models which will dress twelve chairs altogether.

082 BIRD BREAD

It is a bread stand with a birdhouse. Make friends with your winged neighbors - Invite them for a meal. After dining, place this bread stand on the window ledge so birds can feast on the crumbs. Celebrates coexistence and the joy of sharing.

Materials: Wood

* This project exhibited at the La Triennale di Milano Museum during the Milan fair in 2005 as well as in Berlin and Copenhagen.

084 SHADYLACE

Commissioned by Lille2004, Cultural capital of Europe in collaboration with Droogdesign, this parasol was developed together with local lace industries around Lille. It was meant to provide the visitors of the festival with shade and wonder. Its shade out of lace casts shadows of branches and leafs as if you are sitting under a tree.

Descriptions: Page:

DOUBLE CHAIR MICRHOTEL 086

A double chair that holds baggages - A relaxing and comfortable double chair in which to spend the brief or long wait for today's modern day airport mobility. A prompt and cost effective solution with a sleep convertible feature that is also suitable for passengers carrying bulky baggage.

VD 003 / BIKE RACK 088

Six bike VS one car. This bike rack stands on a parc place.

Art direction by Frank Torres.

Materials: Galvanised Steel

Title
I.V. DRIP FEED PLANT POT
Dimension
POT SIZE : 53 X 53 X 30 CM
(HEIGHT WITH DRIP STAND
FULLY EXTENDED 122 CM)
Design / Client / Manufacture
VITAMIN
Quantity
2000
Year
2005

> See also:
Page 056-057

Title
NIKLAS
Dimension
25 X 30 CM DIAMETER
Design / Photo
STERNFORM PRODUKTGESTALTUNG
Client
PROTOTYPE
Year
2004

> See also:
Page 028-029, 034-037, 216-217, 240-241

Title
CANDLESTICKMAKER 1
VASEMAKER 2

Dimension
7.5 X 10 CM 1
6.5 X 15 CM 2

Design / Client / Manufacture
DESIGNFENZIDER

Photo
DESIGNFENZIDER, STUDIO

Quantity
MASS PRODUCTION

Collection
PART OF THE TEL AVIV
MUSEUM OF ART COLLECTION 2

Year
2000 1
2005 2

> See also:
Page 013

2

Title
WATERPROOF 1
COLA CUP 2, 3
GOOD WATER CUP 2, 4

Dimension
27 X 8 X 37 CM 1
9 X 12 CM 2, 3
7.5 X 10 CM 2, 4

Design / Client / Manufacture
QUBUS DESIGN STUDIO

Photo
MAREK NOVOTNY

Quantity
2000 1
250 2, 3
300 2, 4

Year
2001

> See also:
Page 015-019, 021, 030

Title
COLA CUP
Dimension
9 X 12 CM
Design / Client / Manufacture
QUBUS DESIGN STUDIO
Photo
MAREK NOVOTNY
Quantity
250
Year
2001

> See also:
Page 014-015,
017-019, 021, 030

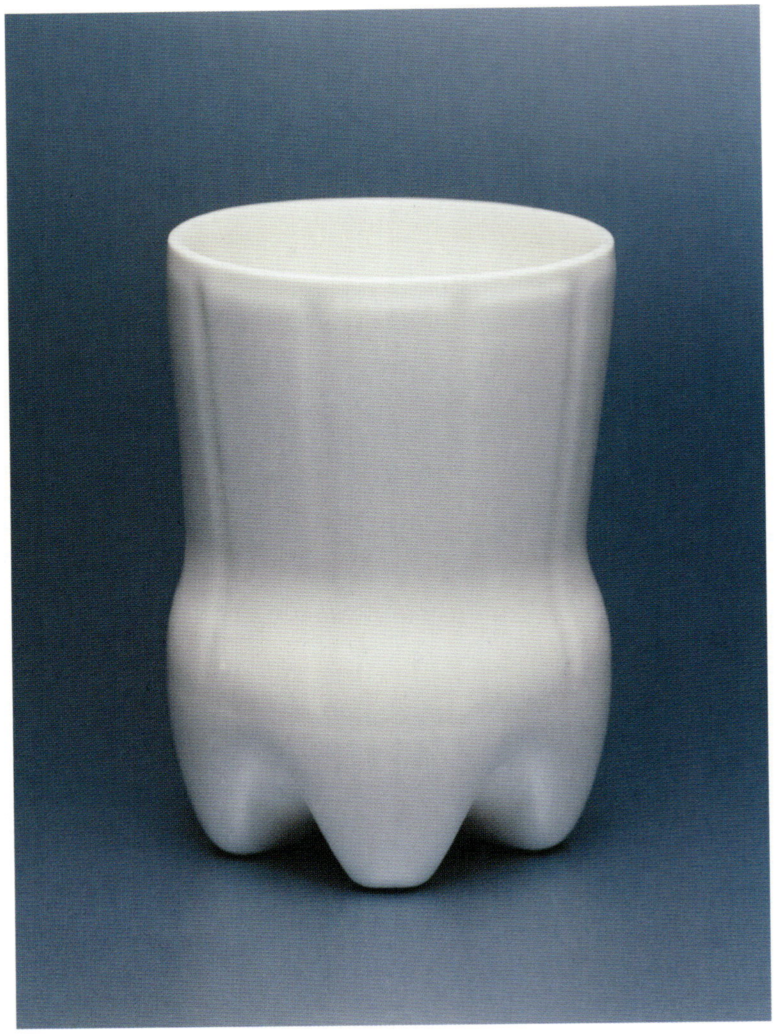

SLUŠÍ!
NOVÁ CHUŤ JI
Coca-Cola
light
0,5
2L

קוקה-קולה

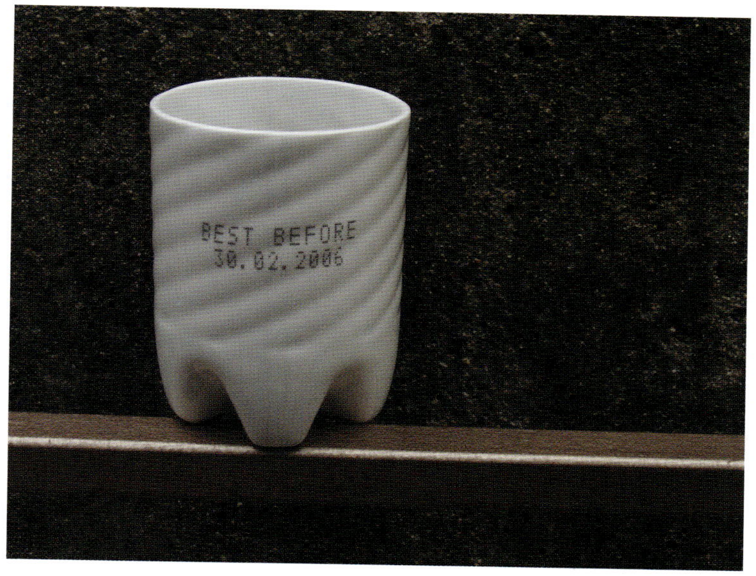

Title
GOOD WATER CUP
Dimension
7.5 X 10 CM
Design / Client / Manufacture
QUBUS DESIGN STUDIO
Photo
MAREK NOVOTNY
Quantity
300
Year
2001

> See also:
Page 014-018, 021, 030

Title

POWER–CUT SOLUTIONS :
E27 CANDLE 1
POWER–CUT PLUG 2

Dimension

11 X 6 X 6 CM 1
12.5 X 9 X 7 CM 2

Design / Photo

MIXKO

Year

2002

> See also:
Page 026-027, 094-095,
153, 166-168

1

2

1

2

Title
ICON CANDLES :
COKE CANDLE 1
WATER CANDLE 1
MAY DAY CANDLE 2

Dimension
6 X 24 CM 1
6 X 21 CM 1

Design / Client / Manufacture
QUBUS DESIGN STUDIO

Photo
ONDREJ HOST

Quantity
845

Year
2004

> See also:
Page 014-019, 030

Title
STAK 1
MORPHSCAPE 2

Dimension
20 X 25.5 CM 1

Design / Photo
KARIM RASHID INC.

Client / Manufacture
DANESE 1
GAIA & GINO 2

Awards
2005 DESIGN PLUS AWARD WINNER 2

Year
2003 1
2005 2

> See also:
Page 023

1

2

Title
DISH NO.3
Dimension
50 X 20 CM
Design
VLAEMSCH ( )
Client
LICENSED BVBA
Year
2004

> See also:
Page 050-051, 112, 119

1

Title
BASKET BOWL 1
FOOTBOWL 2

Dimension
15 X 23 X 23 CM, 66 DIAMETER 1
10 X 21 X 21 CM, 65 DIAMETER 2

Design
MIXKO

Client / Manufacture / Collection
SUCK UK

Quantity
3000 (FIRST PRODUCTION RUN)

Year
2004, 2006 (SUCK UK VERSION) 1
2001, 2006 (SUCK UK VERSION) 2

> See also:
Page 020, 094-095, 153, 166-168

2

Title
JAMES
Dimension
27 CM DIAMETER
Design / Photo
STERNFORM
PRODUKTGESTALTUNG
Client
PROTOTYPE
Year
2004

> See also:
Page 010-011, 034-037, 216-217, 240-241

Title
SOMMELIER
Dimension
7 X 17 CM
Design / Client / Manufacture
QUBUS DESIGN STUDIO
Photo
MAREK NOVOTNY
Quantity
900
Year
2005

> See also:
Page 014-019, 021

Title
OH! MY GLASS
Dimension
8 X 13.5 CM
Design
PROPAGANDIST
COMPANY LIMITED
Photo
CHAIYUT PLYPETCH
Client
PROPAGANDA
Manufacturer
PPP PLASTIC CO.,LTD
Quantity
10,000
Year
2000

Title
UNDINE
Dimension
15 X 15 X 30 CM
Design
BKM DESIGN WORKING GROUP
Photo
GERHARD RAMSEBNER
Client
VIENNA WATERWORKS
Manufacturer
J&L LOBMEYR
Awards
2ND PRIZE,
VIENNA PITCHER DESIGN
COMPETITION 04.2005
Year
2005

> See also:
Page 032, 076-077

Title
TWIN
Dimension
15 CM DIAMETER
Design / Photo
STERNFORM PRODUKTGESTALTUNG
Client
PROTOTYPE
Year
2002

> See also:
Page 010-011, 028-029, 035-037, 216-217, 240-241

V

Voll
milch

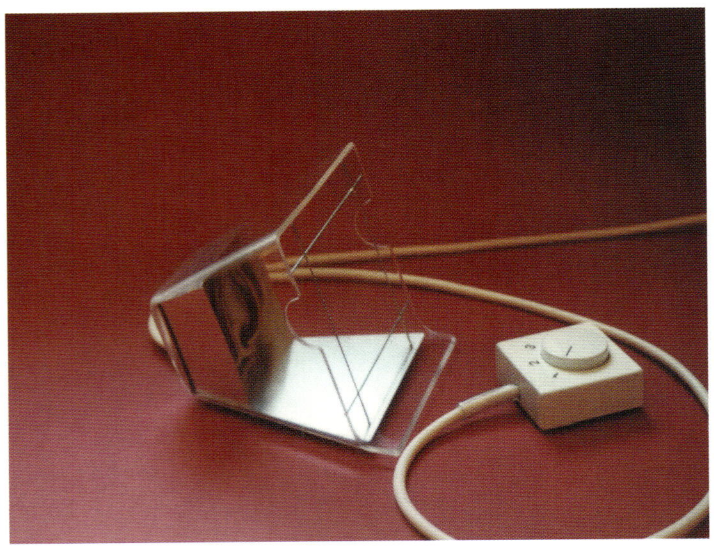

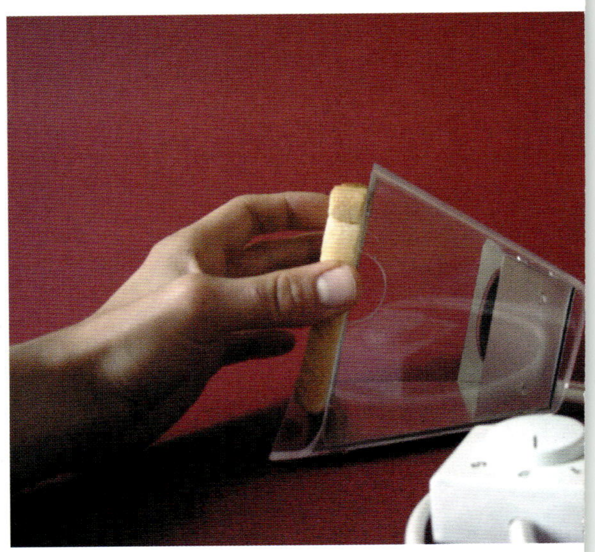

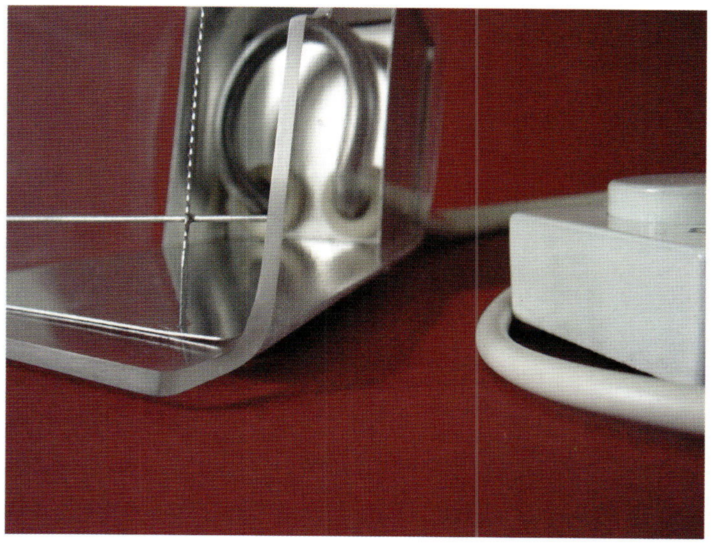

Title
GLASS TOASTER
Dimension
14 X 14 X 18 CM
Design
STERNFORM PRODUKTGESTALTUNG
Client
PROTOTYPE
Awards
1ST PRIZE, NACHLUX DESIGN AWARD 2001
Year
2001

> See also:
Page 010-011, 028-029, 034-036, 216-217, 240-241

Title
BUON APPETITO
Dimension
DUVET : 135 X 200 CM
PILLOW : 80 X 80 CM
Design / Client / Manufacturer
MISS GESCHICK & LADY LAPSUS
Photo
NATHALIE MOHADJER,
LAURENTIUS SCHMEIER,
ALEXANDER LEMBKE
Quantity
300
Year
2006

> See also:
Page 039

Title
PASTA FORK
Dimension
2 X 2 X 15 CM
Design / Photo
TIM PARSONS
Client
COMPETITION ENTRY
FOR BARILLA PASTA
Quantity
PROTOTYPE
Year
2002

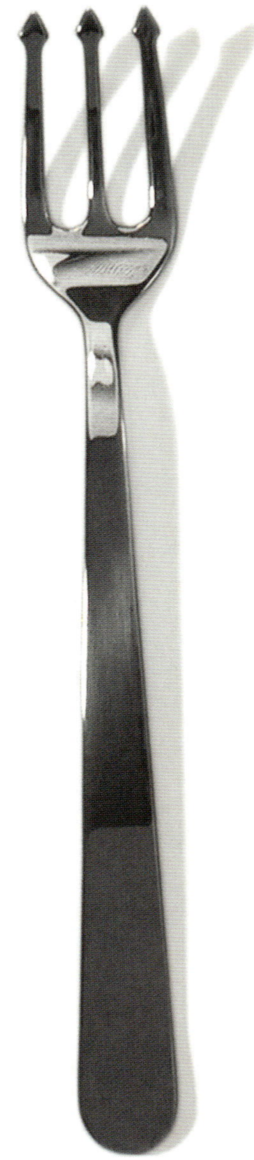

Title

MP SPOON 1
DB SPOON 2

Dimension

2.5 X 11.5 X 1.5 CM

Design / Photo

ONTWERPWERK
MULTIDISCIPLINARY DESIGN

Client

PRIME MINISTER BALKENENDE /
MINISTRY OF FOREIGN AFFAIRS 1
DROOG DESIGN 2

Manufacture

ONTWERPWERK,
MUTIDISCIPLINARY DESIGN 1
DROOG DESIGN 2

Quantity

140 1
IN PRODUCTION 2

Year

2004 1
2005 2

1

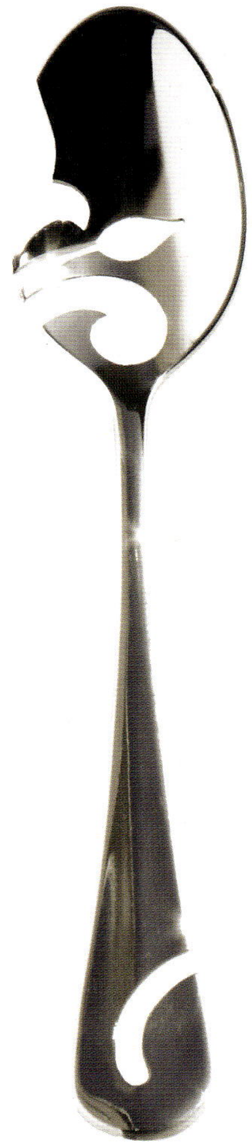

2

Title
BRUSH COUPLE 1
MATCHES 2
Dimension
30 X 10 X 9 CM 1
VARIOUS 2
Design
ADRIEN ROVERO STUDIO / ECAL
Photo
FELIX IMHOF 2
Client
DIM, DIE IMAGINÄRE MANUFAKTUR 1
ADRIEN ROVERO 2
Quantity
PROTOTYPE 2
Year
2004 1
2002 2

> See also:
Page 043, 045, 088, 096-097, 102, 238-239

1

2

Title
WOOFER
Dimension
55 X 45 X 25 CM
Design
BURO VORMKRIJGERS
Photo
NIELS VAN VEEN
Manufacture / Collection
CULTIVATE
Year
2006

> See also:
Page 120, 122-123, 135,
194-195, 218-219

Title
ROLLING
Dimension
SMALL ROUND : 16 X 12 CM DIAMETER
MIDDLE TRIANGLE : 27 X 27 X 16 CM
HIGH SQUARE : 30 X 30 X 42 CM
Design / Client / Manufacturer
ADRIEN ROVERO STUDIO
Photo
PHILIPPE JARRIGEON
Quantity
PROTOTYPE
Year
2006

> See also:
Page 042-043, 088,
096-097, 102, 238-239

Title
OH CHINA
Dimension
VARIOUS,
27 X 11 CM DIAMETER
Design / Client / Manufacturer
FIFTY-EIGHT B
Photo
CECIL MATHIEU
Quantity
50
Year
2006

> See also:
Page 047, 160-161

Title
GHOST TREE
Dimension
45.7 X 182.9 CM
Design / Manufacturer
ERICH GINDER STUDIO
Photo
NEIL LUKAS
Year
2004

> See also:
Page 049

Title
MOOSE 1
ROEDEER 2
DEER 3
Dimension
96.3 X 61.8 X 86.2 CM 1
12.1 X 17.6 X 30 CM 2
52 X 25.5 X 70 CM 3
Design
VLAEMSCH ()
Client
LICENSED BVBA
Year
2004

> See also:
Page 024-025, 051, 112, 119

1

2

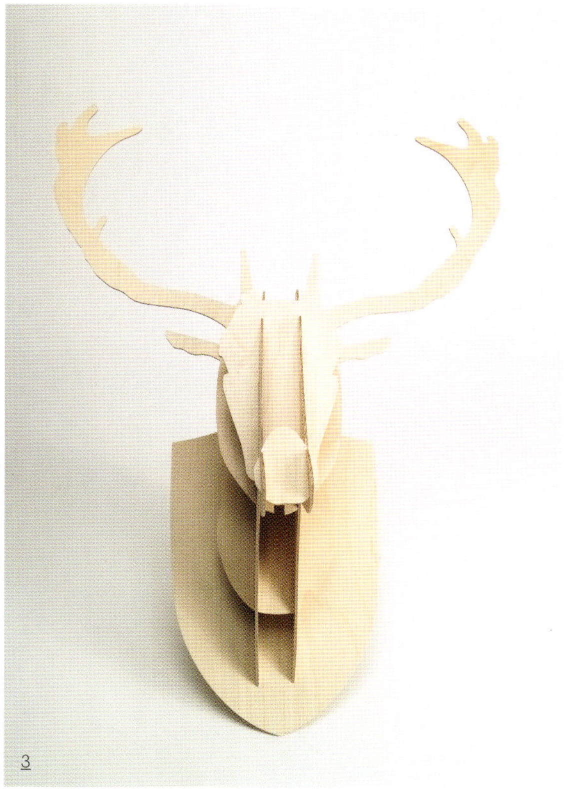

3

Title
HEATWAVE
Dimension
55 X 70 X 10 CM PER ELEMENT
Design
JORIS LAARMAN STUDIO
Photo
BAS HELBERS
Client
DROOG B.V. & JAGA
Manufacture
JAGA
Collection
DROOGDESIGN, MUSEUM BOYMANS VAN BEUNINGEN, FRAC NORD PAS DE CALLAIS, KUNSTPAVILJOEN
Awards
RED DOT DESIGN AWARD 2006,
IMM INTERIOR INNOVATION AWARD 2004
Year
2006

> See also:
Page 052, 246-247

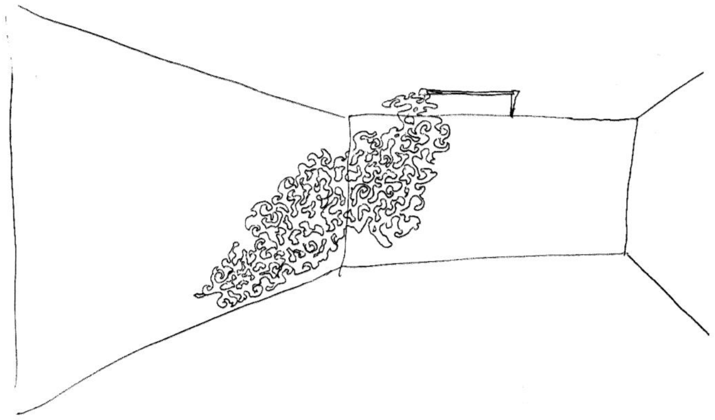

Title
ROADKILL CARPET
Dimension
165 X 225 CM
Design / Photo / Manufacture
KARIN VAN LIESHOUT
Quantity
PROTOTYPE
Year
2001

> See alsо:
Page 054, 090-091, 184-185

Title
SPRUCE
— SWEEPER LIGHT
Dimension
30 X 30 X 180 CM
Design / Client / Manufacture
VITAMIN
Photo
GRANT SQUIBB
Quantity
50
Year
2005

> See also:
Page 009, 057

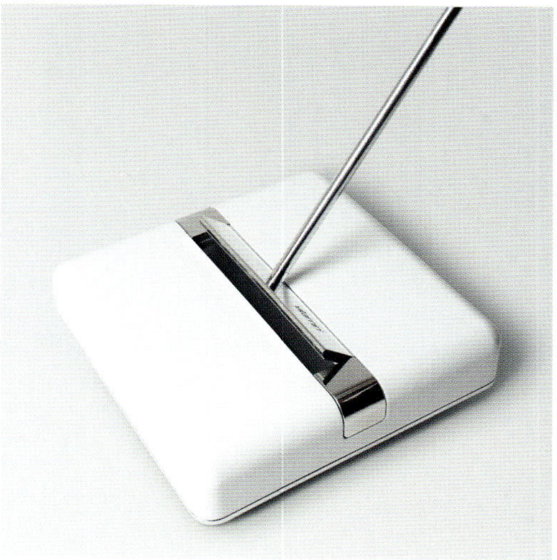

Title
STICKY LAMP
Dimension
8 X 18 X 25 CM DIAMETER
Design
CHRIS KABEL DESIGNS
Photo
CHRIS KABEL STUDIO, FERRY EKEL
Manufacturer / Collection
DROOGDESIGN
Year
2001

> See also:
Page 059, 062-063, 084-085, 182-183

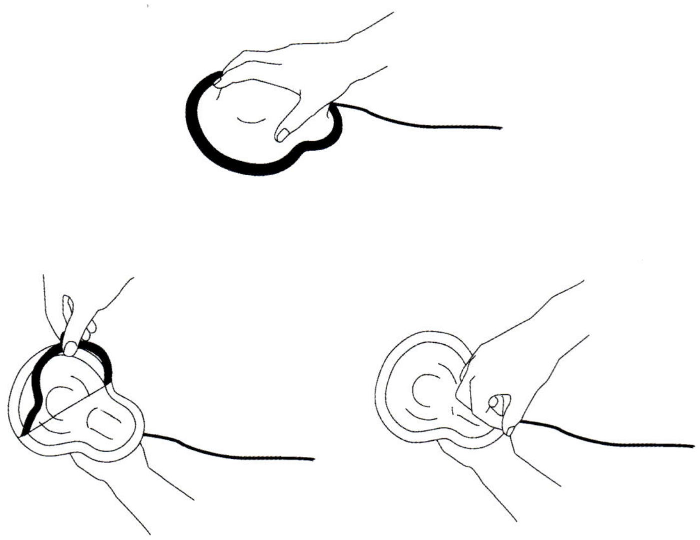

PHILIPS

Title
SNOTTEBEL
Dimension
EACH BRANCH 200 CM
Design / Manufacturer
SYLVAIN WILLENZ
DESIGN STUDIO
Photo
JULIEN DE WILDE
Quantity
LIMITED EDITIONS
Awards
SELECTED IN HENRY
VAN DE VELDE AWARDS,
DESIGN FLANDERS,
BELGIUM 2004
Year
2005

> See also:
Page 061

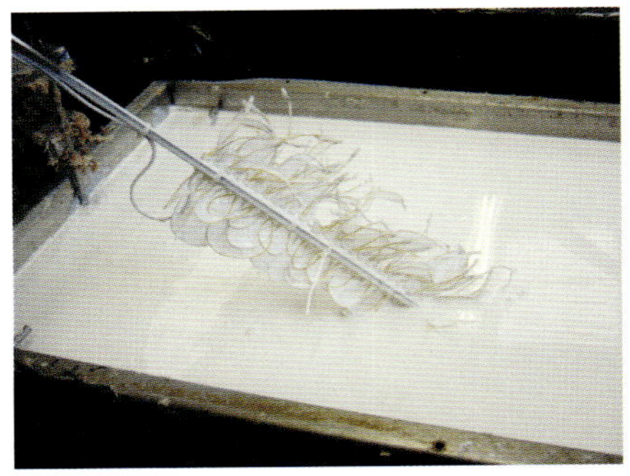

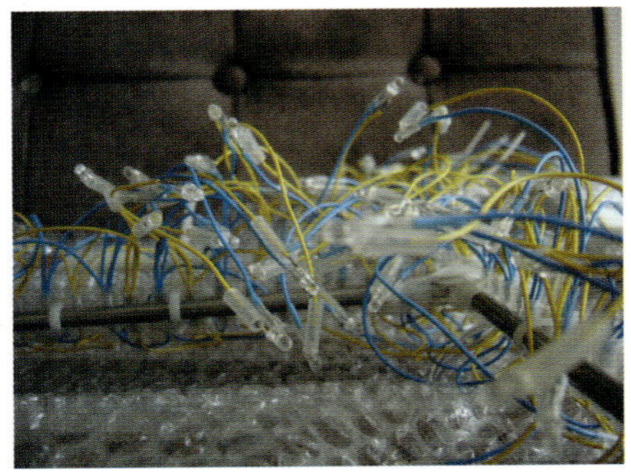
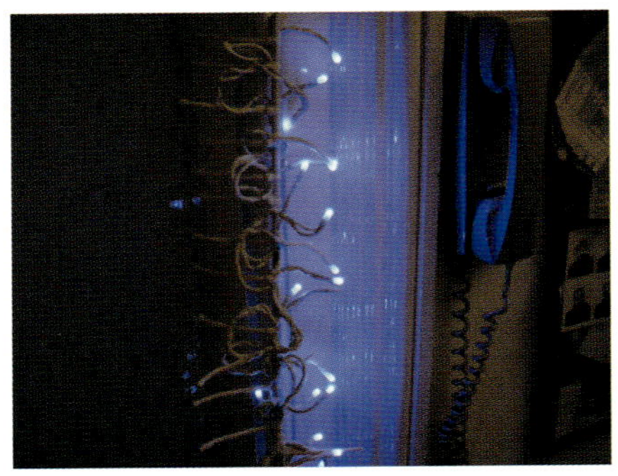

Title
1TOTREE
Dimension
14 X 14 X 14 CM
Design
CHRIS KABEL DESIGNS
Photo
CHRIS KABEL STUDIO
Manufacturer
LIMITED BATCH
Quantity
97
Collection
STEDELIJK MUSEUM AMSTERDAM DESIGN COLLECTION, HANGARAM DESIGN MUSEUM, SEOUL, SOUTH KOREA

Awards
NOMINATED FOR THE ROTTERDAM DESIGN PRIZE
Year
2002

> See also:
Page 058-059, 063, 084-085, 182-183

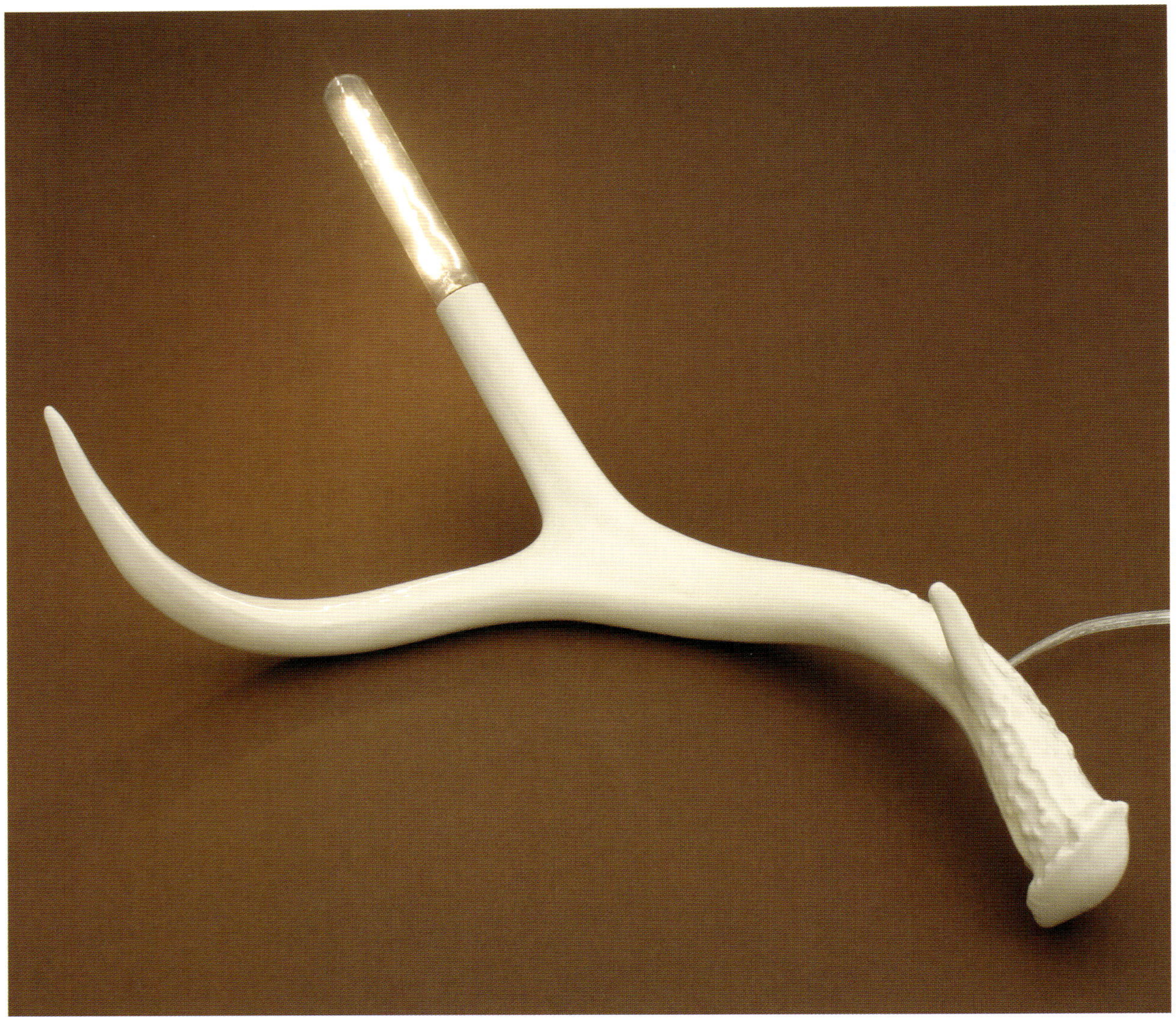

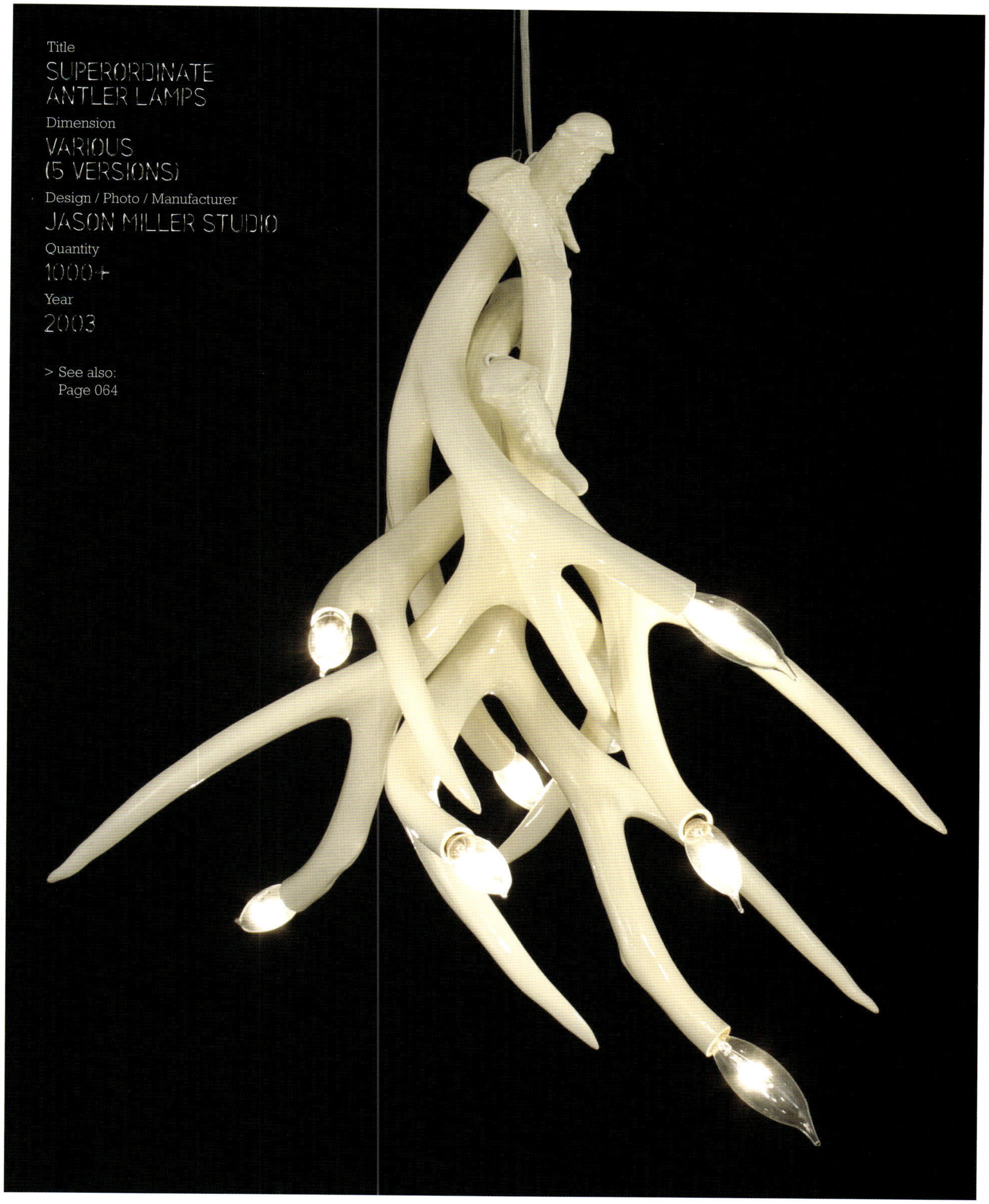

Title
SUPERORDINATE ANTLER LAMPS
Dimension
VARIOUS (5 VERSIONS)
Design / Photo / Manufacturer
JASON MILLER STUDIO
Quantity
1000+
Year
2003

> See also:
Page 064

Title
SPAGHETTI CHANDELIER

Dimension
950 X 650 CM (LARGE, 25 BULBS)
600 X 550 CM (SMALL, 15 BULBS)

Design / Client / Manufacture
PARK PLUS

Photo
YUN SUKMU

Year
2005

Title
ELECTRIC SPAGHETTI PRINCESS CHANDELIER
Dimension
APPROX. 100 X 150 CM (LARGE, 8 BULBS)
APPROX. 65 X 150 CM (SMALL, 6 BULBS)
Design / Photo / Manufacturer
STUDIOMOLD
Year
2006

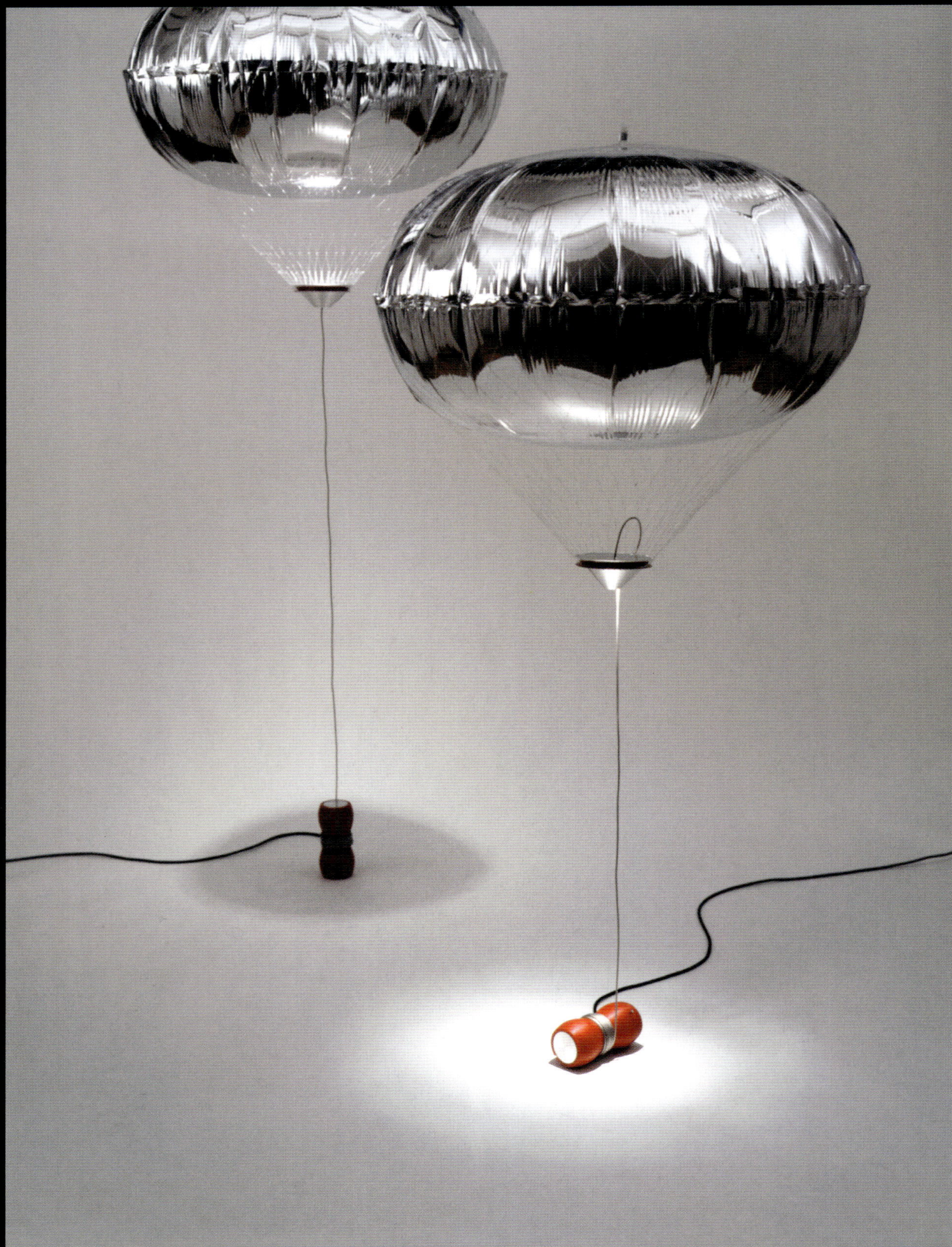

Title
WORLD VIEW
Dimension
100 X 100 CM
(HEIGHT UP TO 500 CM)
Design / Manufacture
AFRODITI KRASSA LTD
Photo
FRANK THURSTON
Client / Collection
THE DIVINE COLLECTION
Quantity
20
Year
2002

> See also:
Page 068, 169, 200-201, 207

Title
POODLE
Dimension
75 X 50 X 50 CM
Design / Client / Manufacture
STUDIO VAN EIJK &
VAN DER LUBBE
Photo
ELI (POODLE MILAAN)
Year
2002

> See also:
Page 071, 116-117, 158-159

Title
BATH & BEYOND
Dimension
70 X 58 X 70 CM
Design / Photo / Client / Manufacture
REDDISH
Quantity
LIMITED EDITIONS
Collection
PRIVATE COLLECTION
Year
2002

> See also:
Page 073, 082-083, 189

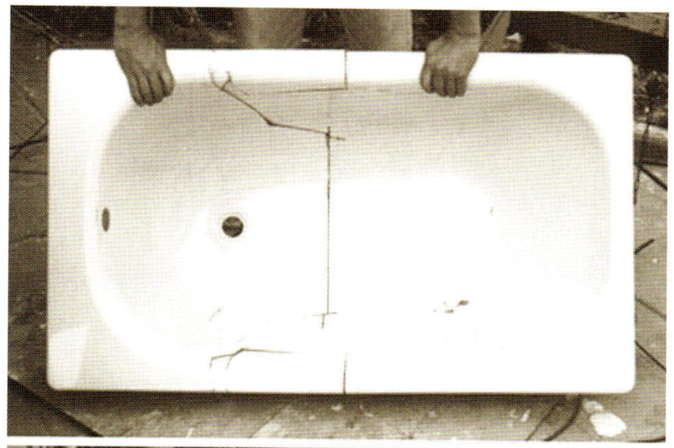

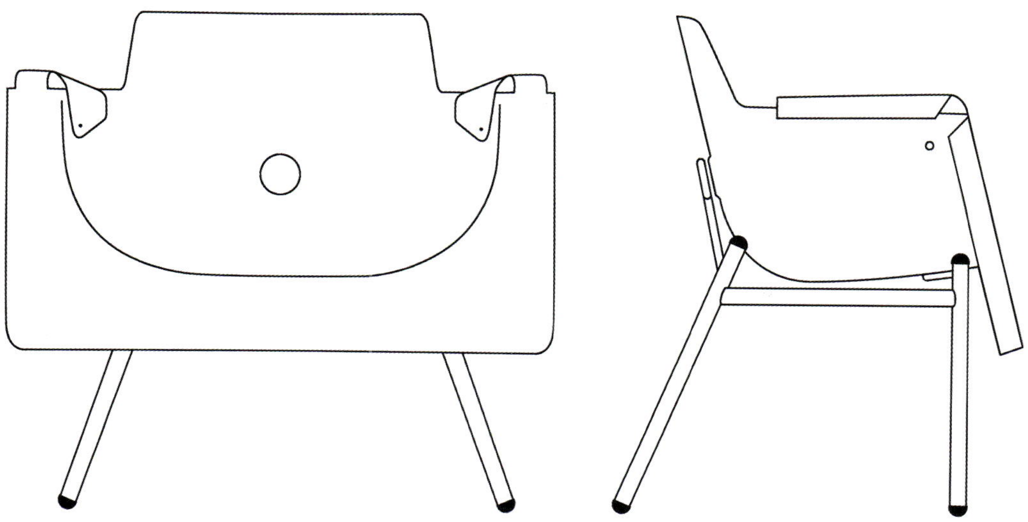

Title
BROOMSTOOL
(OLD NAME BESENHOCKER)
Dimension
50 X 35 X 50 CM
Design / Client / Manufacture
OLSCHEWSKI
PRODUKTDESIGN
Photo
MIRIAM HOLZ
Quantity
MAX. 50 PER YEAR
Year
2005

> See also:
Page 075

Title
BATHROOM (OLD NAME DIE OFFENE GESTALTUNG)
Dimension
TOGETHER APPROX. 240 X 180 X 80 CM
Design / Manufacture
OLSCHEWSKI PRODUKTDESIGN
Photo
MIRIAM HOLZ
Client
DIPLOMA
Quantity
12
Year
2006

> See also:
Page 074

Title
FALB
Dimension
40 X 50 X 100 CM
Design / Manufacture
BKM DESIGN WORKING GROUP
Photo
GERHARD RAMSEBNER
Quantity
SMALL SERIES

Awards
1ST PRIZE, CAIAZZA MEMORIAL CHALLENGE 2004 INTERNATIONAL DESIGN COMPETITION, UDINE, ITALY
Year
2004

> See also:
Page 032-033, 077

1

Title
BIG SOCKS 1
PICK POCKET 2
Dimension
40 X 40 X 80 CM 2
Design / Client / Manufacture
DRESSING
Photo
ANGELA FRANKE
Quantity
SMALL SERIES
Year
2003

> See also:
Page 079-081

Title
COSY COVER
Dimension
120 X 120 X 80 CM
Design / Client / Manufacture
DRESSING
Photo
ANGELA FRANKE
Quantity
SMALL SERIES
Year
2003

> See also:
Page 078-079, 081, 118

Title
BIRD BREAD
Dimension
17 X 47 X 16 CM
Design / Photo / Client / Manufacture
REDDISH
Quantity
PROTOTYPE
Year
2004

> See also:
Page 072-073, 082, 189

Title
SHADYLACE
Dimension
225 X 215 CM DIAMETER
Design
CHRIS KABEL DESIGNS
Photo
CHRIS KABEL STUDIO,
DROOGDESIGN,
DANIEL KLAPSING
Manufacture
SYWAWA (BELGIUM)
Collection
DROOGDESIGN,
HANGARAM DESIGN MUSEUM,
SEOUL, SOUTH KOREA,
BOIJMANS VAN BEUNINGEN
MUSEUM ROTTERDAM
Year
2004

> See also:
Page 058-059, 062-063,
085, 182-183

Title
DOUBLE CHAIR MICRHOTEL

Dimension
90 X 100 X 120 CM

Design
STEFANO CASSETTI DESIGN OFFICE

Client
JOINT VENTURE : BRATISLAVA–BRESCIA –PARIS–MILANO–EU

Manufacture
DESIGN CENTRUM BRATISLAVA

Year
2007

> See also:
Page 087

Title
VD 003 / BIKE RACK
Dimension
340 X 170 X 45 CM
Design
ADRIEN ROVERO STUDIO
Photo
IANN GROSS,
EMILIE MÜLLER
Client
INOUT
Quantity
PROTOTYPE
Year
2006

> **See also:**
Page 042-043, 045,
096-097, 102, 238-239

FUN + FUNCTION

SCHILLER, ONE OF THE MOST REVEALING AESTHETIC THINKERS OF THE BEGINNING OF THE 19TH CENTURY, MADE AN ATTEMPT TO BREAK THE BOUNDARY BETWEEN ART AND NON-ART BY INTRODUCING THE IDEA OF PLAY AS AN OVERARCHING TERM. MAN IS ONLY TRULY HUMAN WHEN HE IS AT PLAY, AND THIS INSIGHT, FOR SCHILLER, WOULD SOME DAY "UNDERLIE THE WHOLE EDIFICE OF AESTHETIC ART AND EVEN MORE DIFFICULT ART OF LIFE". THE FOLLOWING DESIGN OBJECTS HAVE THE PURPOSE TO PARTICIPATE IN PEOPLE'S LIFE BY PROVIDING ENJOYMENT. TWO CENTURIES LATER, THESE DESIGN OBJECTS ASK HUMANS TO SMILE BY PLAYING WITH EVERYDAY LIFE.

Page: Descriptions:

089 'ELECTRIC JOEY' TV STAND

The design is an inspiration from life and social comments.

Joey the family budgerigar (small bird) would sing all day, a constant background noise. He would sleep only in the dark. We treated him as part of the family.

Simple unobtrusive TV stand with cover for 14-inch televisions suitable for rooms where the focus point is not always the television.

090 WOODEN USB STICK

USB memory sticks are always nice to have around when transferring files from computer to computer.

These sticks are made to stand out against its natural environment of computers and offices. They are picked from the woods and are manually selected on their natural beauty, professionally handmade into unique and personal USB memory sticks.

092 UNIVERSAL CONNECTIONS

What is the difference between analogue and digital, physical and virtual? What connects and bridges these apparently different worlds? USB is nowadays used as a synonym for an universal connection. Originally only used by computer experts and nerds, this so-called former high-tech standard has developed into an everyday item and low-tech symbol for such universal connections. This is an example of something democratic evolves and develops and therefore often even receives a totally different usage than was originally intended. Given the current, it would even be argued that there is a 'USB subculture'!

094 EARPHONE SPEAKERS

Enlarged earphone speakers produced to look and sound great either hung from the ceiling or placed on the floor.

Descriptions: Page:

MY TURN-ONS 095

As well as experimenting with scale to take objects out of context, the designer also like to experiment with objects in their context. Looking at the electric plug; each of its elements is questioned: the plug itself, the flex, the wall socket and its graphics. New ideas started to emerge by making simple changes. The ideas that didn't quite work spawned more ideas, and the stupid ones lead to better things.

SWAROVSKI (PENDELUM LIGHT, TAPE IT!) 096

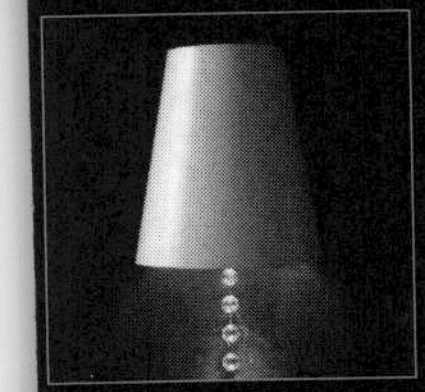

Work with the crystal brand at Ecal. Two purposes: Tape it & Pendelum light.

Materials: Crystals & Miscellaneous

18KT GOLD BIC PEN LID, GOLD PILLS 098

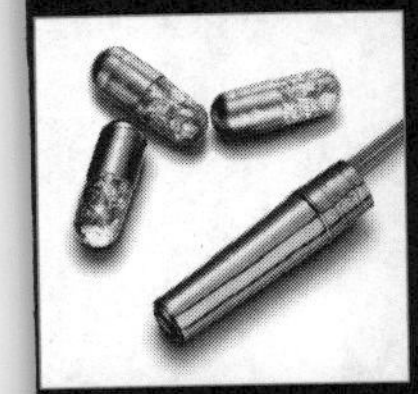

The pen cap is a classic icon of great design. But in drug culture it is also a tool to snort cocaine with.

The gelatin cases of these pills are filled with gold leaf. 'All you have to do is swallow the pill and 24 hours later your shit turns gold,' says Wong.

ULTIMATESETTING AKA KILLERRING 098

The diamond in this ring is upside down so that the sharp side is exposed. 'It takes protection into your own hands,' says Wong. As an engagement ring, it is not only a token promise of love but also doubles as a protection weapon when you are not there – a minimum of1kt diamond is powerful enough to cut through the skin down to the bone.

Page: Descriptions:

099 **SMOKINGMITTENS**

The nylon mitten has a rivet hole inserted into it. 'This idea was inspired by the smoking ban in bars in New York. People were complaining about going out in the cold to smoke so I designed this to allow them to smoke and keep warm,' says Wong. They come as a pair but only one of them is designed with a hole as they are left-right reversible.

100 **LAST BUNCH OF FLOWERS FOR MY GIRLFRIEND (CAL .357 MAGNUM)**

Set of 5, limited edition (250 sets), for sale. Cartridge out of metal, bullet in solid sterling silver.

101 **DUTCH DELIGHT**

Holland is known as an open-minded country when it comes to drugs and sex. The sextoys they designed are not just a joke. They refer to the liberated sexual mentality of the Dutch people. Making a dildo out of porcelain is also a good example of using the right material for the right product. Glazed porcelain has a soft and smooth skin which is pleasant to touch and easy to clean. The hollow shape can be filled with hot water or crushed ice. The cork is used to close it off. Decorated with Delfts' Blue motives, it becomes more than just a cheap sex toy for sale.

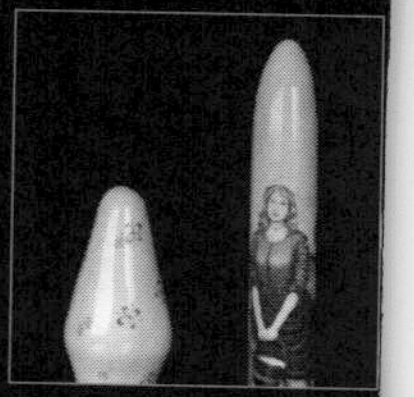

102 **DIP 1/3**

It's a series of traditional silver Christofle pieces dipped into liquid rubber. This red rubber is a process usually used for the tools in the industry world. The confrontation between luxury crafted silver pieces and a rough industry process gives contemporary pieces. It has been done during a workshop at Ecal with the collaboration of Christofle.

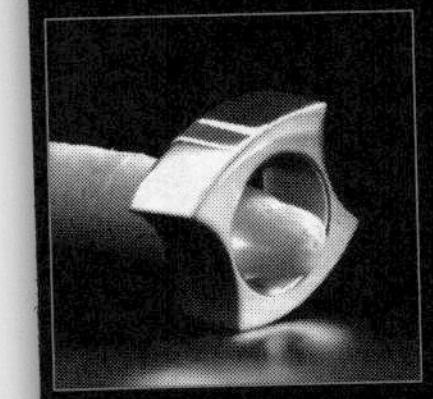

THE SUN, THE EARTH, THE MOON, THE KISS

103

Here are four silver rings of which three, 'The Sun', 'The Earth', and 'The Moon', belong to the same series. 'The Kiss' is made in the same technique and style but is a separate work.

The rings are hollow, constructed from sheet metal and therefore fairly lightweight even though they look very massive. Also, they are ergonomically designed so they fit very comfortably on the fingers next to the one its worn on, so they are easy to wear.

TANK YOU

104

'Tank you' is a vase made in 'porcelaine de Limoges'. All vases are handmade and each model is unique. Tank you is a love story between a tank and a flower. Hold your guns, this is flower power.

It is available in 5 versions: white, flowers, platinum, gold plated and think tank.

PROAESTHETICS

106

An old man, by most people's standards, broke his hip in his late seventies, but with all the dignity worthy of his age and experience intact. Upon being able to walk again he was given a plastic crutch, which the designer thought was rather incompatible with this person. Once he was able to walk without the aid of the crutch, it was left abandoned in a corner, as a lingering memory of his misfortune. With this in mind the designer set about redesigning medical prosthesis in the hope of arriving at more dignified solutions. The exploration lay not so much in the chosen forms, as they are largely dictated by the human body, but rather in a change of material: porcelain. A fragile yet strong material, hygienic whilst elegant, in other words with all the right paradoxical sought after qualities.

The result, perhaps more poetic than practical, does however reflect the inherent beauty of recovery and mirrors the healing process of our mending bones. Slightly awkward and very fragile, just like us, they cry out, 'handle with care'! Upon recovery, disposing of these trustworthy companies is simply not an option. Instead, they can be kept and treasured, and exhibited amongst our finest (bone) china.

Page: Descriptions:

110 HEAD HAND BAG

This project explores our ability to translate an object into a story and a story into a product. The biblical story about Judith and Holofernes and its visual representation from the Renaissance were inspiration for the design of the bag. The bag enables its users to experience a sense of victory without the need for violent action.

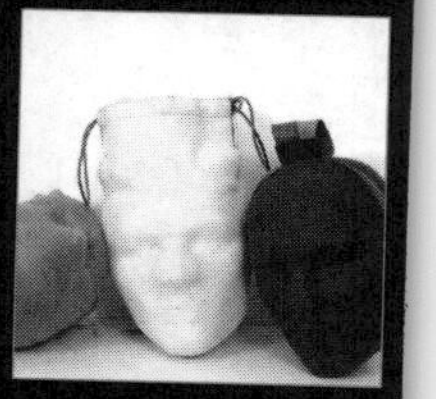

112 INTERNAL WATCH

The Internal Watch is encased in leather, rendering it non-functional as a timepiece. The hidden watch transcends typical associations of status by isolating the emerging form rather than acting as a flashy signifier. The Internal Watch is available in black, white or natural leather.

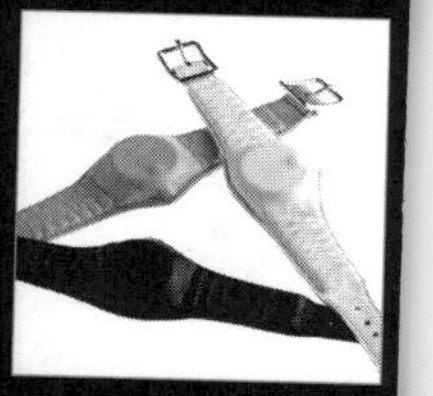

113 CHARACTERIZED DESIGN — ICON WATCH

The conception of 'Characterized Design' means extracting the essential qualities of things, restructuring them. The watch is designed with no border between graphic and product; and it is made by dots.

114 PEEPSHOW

'Peepshow', where all beauty is hidden on the inside. For your eyes only. Anywhere you go you see safety camera's registrating life, people get filmed on every corner of the street. You get the feeling that maybe you have got something to hide. By carrying a 'Peepshow' bag you do have something to hide. Let the camera's film you, they will never reveal the images inside your bag!

Descriptions: Page:

GUARDIAN ANGEL 114

'Guardian Angel', makes a strong statement about growing rates of crime and violence. Each of the bags appears to contain a gun or a knife. The idea grew out of the increasing number of stories in the news that focused on lawlessness and antisocial behaviour. People's perceptions and the fear of violence soon grew to be greater than the number of crimes actually committed. As a response, Vlieger & Vandam wanted to give people the impression of feeling secure, arming them with a 'Guardian Angel' handbag.

NOT-A-LAPTOP 115

'Not-a-laptop' is a laptop bag, disguised as a shopping bag full of groceries. Now your laptop is safe from being stolen.

Black laptop bag made of textile covered polyethylene foam. The front shows a 3D sculpture of a grocery bag: vegetables, a bottle of liqueur and a carton of milk. This bag can be carried as a tote bag and also as a rucksack. Open the zip-closure at the back to store your laptop, papers, pens and business cards. Small zipped pocket at the top of the bag is designed to store your keys and wallet.

HUNGURY JACK 115

'Hungry Jack' loves your rubbish and he is hungry for more.

These bags have an outline of a Dachshund: smooth-haired on one side and an X-ray photograph on the other side.

Are you curious what is it inside his belly?

Bag with front made of black Pony hair. There is a silk-screened image on leather on the other side. Shoulder strap, zip closure and small inner zipped pocket.

LADY'S BAG ME AND MY BERETTA 116

Lady's Bag – You never know what's in it. The design is a reaction to changing society nowadays.

Page: Descriptions:

118 HIDE & SHOW

In everyday life clothes are spread out in the room or disappear behind closet doors. 'Hide & Show' transfers colours and aesthetics of fabrics into an element of interior design. Like a frame the furniture embraces the garments thus emphasizing the colour play. At the same time it is hiding the chaos and protecting clothes. Depending on season or personal preferences the cabinet changes its appearance – therefore no closet is like another.

Material: Wood plate, Beech, Coated

119 BOOTBAG

Born in 1986, Saskia Marcotti is probably the youngest Belgian designer in history to have had a product in production. BootBag is a perfect example of good design and combines a minimum of effort with a maximum effect. Simply by making a hole in a rubber boot, the boot becomes a most original bag – even the dullest looking person instantly becomes someone attractive and funny. BootBag kicks the ass of all other bags!

120 ABOUT TIME

Poetry in motion; this clock reveals the time by rolling around your desk, and tells you the time in one long continuous sentence. Designed in reaction to our stressed life's, where we tend to plan our daily life to the minute, this clock only tells you 'it's about six o'clock' or 'it's almost seven now'. While rolling around your desk, the slow but constant almost meditative motion allows you to relax, maybe even forget about time for a few minutes.

121 SEEING DOUBLES

By using the red and green transparent films on the clock hands, we are able to display both the minute and hour readings on the same set of markers. The design of Seeing Doubles, in effect, represents time in the manner of a digital clock. But of course, all is done with the classic analogue dial. In a time when many have embraced for new technologies; we were impressed by the fact that even with such simple use of the colour theory / visual trickery we were able to create something that is new, exciting. The final product certainly makes clock-watching so much more interesting.

PONG CLOCK

122

In the old school Pong TV game you could put the paddles in such a position, that the ball kept bouncing forever. Quite hypnotic, and it became the inspiration for this time-killer/clock.

While this game of Pong randomly plays, the so-called 'players' score the time. The left player scores the hours, and the right player scores the minutes, creating an unusual timepiece, ideal for those Friday afternoon moments.

CUCKOU

124

Digital clock, metal housing, chromed nails and chain, cuckoo (the cuckoo died from natural causes in 1958)

FLYING FISH BOWL

126

A giant wall mounted revolving fish bowl allows your pet to travel the polluted ocean. The idea based on the fact that the water stays at the same position so the fish doesn't feel the movement. To satisfy the fish needs in a new way, containers are shaped as industrial buildings that usually cause pollution at the oceans like power station, cities drain, factories, etc. The scenery is constantly changed so it looks like the fish is reaching different spaces rather than staying in the same tank. Containers located around the fish tank would carry the oxygen and the moving decorations would supply oxygen for the fish.

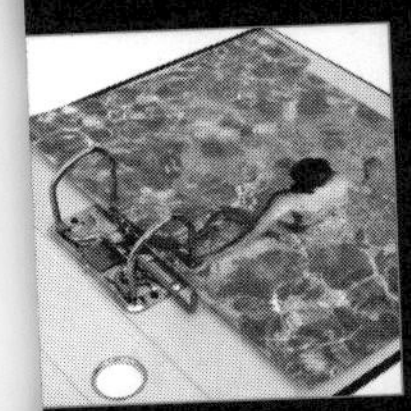

DIP

128

Dip, dive and surface – a little flight away from the daily office routine.

While doing research on the subject of workplace design and stress management, dip was first nothing but a visual gimmick touching on the issue of daydreaming at work. From this a real product emerged whose task is to bring a bit of life and colour into the sometimes dull office existence.

Limited-lot production by StauffacherBenz for the exhibition 'Design? Art?' at the Kunsthaus Langenthal 2004.

Page: Descriptions:

129 ## 16 ANIMALI

Conceived in 1960 for Christmas greetings, this original creation by Enzo Mari underlines a double function: it is a puzzle object and also a 'construction game' that allows a free composition of elements. The 16 ANIMALI and 16 PESCI can be considered a small art sculpture for your home; the elements can be used both individually and in multiple compositions. All of them are recognizable on sight. They fit into one another jigsaw-fashion and can also be stood upright because the vertical position is balanced by the thickness of the precious wood they are made of. There are no given rules.

130 ## WAR BOWL SERIES

A bowl made by melting model soldiers. This version represents the Battle of Waterloo. Half British Artillery, half French Infantry. The designer aimed to make a material that had a voice of its own, not just a passive, lifeless thing. Started with the idea of combining small objects to create something larger, it ended up with using toy soldiers as the ingredients. On one hand they were a fun childhood toys that recalled happy memories and on the other hand, they were a representation of war with its associated horrors. This is a contrast that the designer finds interesting. (www.dominicwilcox.com)

132 ## NAPKING

To entertain and amuse your friends while staying at table, it is what the napking wants to do. More than a vulgar napkin, napking proposes a game, a challenge and even a charade to the one who handles it. It is based on the oriental folding techniques of paper: the origami technique. This napkin is used by all ages: from 6 to 96.

133 ## VOODOO KNIFE BLOCK

Knife block set for daily food rituals. It is daring and useful, with a cutting-edge design. Made of ABS plastic, it holds five stainless steel molybenum vanadium knives (a paring knife, a bread knife, a carving knife and a large and small chopping knife) held in place through an innovative system which incorporates small seemingly invisible magnets. The blades are protected at the rear by transparent plastic sheaths. Originally crafted in blood-red plastic only, it is now available in other colours too. In Europe, Asia and Australia it's called Voodoo, while in the USA, it's distributed under The EX trademark.

Descriptions: Page:

GLASSBULBS 134

The product offers two lives. The user looks forward to the end of one, with the beginning of the other. Once the lightbulb has burnt out, a wineglass comes to life.

CECI 135

Ceci n'est pas une lampe!

A broken bulb can not function, or can it? This lamp plays with the stereotypes of our daily lives which are imprinted in our minds. By clever use of a high power led, lamp becomes shade, and the impossible becomes possible. The bulb is hand-blown from heat resistant Pyrex glass, safe to the touch with no sharp edges. It is numbered and signed by the designer, limited edition.

BAR MAN RING, POCKET BRUSH 136

A tool you need. It is a ring that helps you to open any bottle. Useful as a ring, and allowing you have the possibility to open bottles in a magic way, just by going with the hand through the bottle. It is made with stainless steel.

The pocket brush is an object that takes another identity to be re-contextualized, it means that with the shape of a key it can stay in a pocket, for example, apart from that, the object is completely simple and says nothing. It also has no clear function, that's why, in the official text it says: If you buy it, it cleans your pocket out.

FLAMP 137

Luminous phosphorescent object in the shape of a lamp. It can glows in the dark for 20 minutes.

Page: Descriptions:

138 ANIMALS

Who wouldn't want a horse to lighten up your home and a pig to serve your guests? It is a set of furniture to fall in love at first sight or hate forever.

Material: Polyester / Colour: Black

139 KEBAB LAMP

It is made with a collection of found objects, antiques and other miscellany no longer in use to delight the viewer with an explosion of colour and reference. Built as a totem, it is carefully composed to contain stories and meanings amongst the eclectic objects they include and consequently each lamp is unique and filled with pieces from different eras that allude to the constant turning of fashion and style. The concept behind is the way people relate to and value objects. It plays with this sense of value as fine porcelain and plastic come together to form a sculptural composition with its own integrity.

140 DISPOSABLE CHANDELIER

It is created from 416 disposable plastic wine glasses with a pink fluorescent light source.

141 MILLENNIUM CHANDELIER

The chandelier is made of 1000 exploded Party Poppers collected (on 01.01.00) after the Millennium celebrations in London. Each Popper is suspended on a line from a platform above. The sculptural shape sways and moves like an organic form when hit by a breeze of air. Although this original light is a 'one off' piece due to the narrative it will be recreated using new party poppers.

Descriptions: Page:

TIDE CHANDELIER 142

The Tide chandelier is part of a larger body of work based on the collection of 'man made' debris washed up on a specific stretch of Kent coastline. The designer has been collecting material over a 2-year-period and the work is still in progress. The material collected is sorted and categorized and several individual pieces of work were produced. It is created from clear and translucent objects primarily made of plastic. Each object is different in shape and form, yet they come together to produce one sphere. The sphere is an analogy for the moon that affects the tides, which in turn wash up the debris.

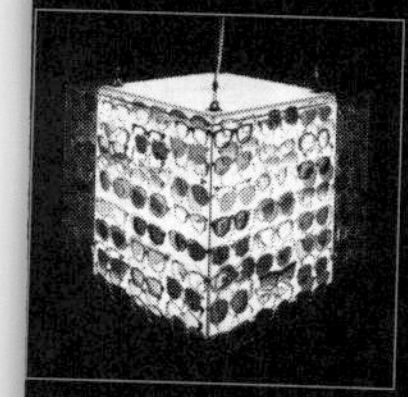

TWENTYTWENTY 144

Twentytwenty is a light created from hundreds of prescription spectacles sourced from a charitable organization that redistribute discarded glasses to third world countries. By salvaging the glasses an elegant and analogous line has been made between their old and new purposes - from prescriptive items for improving sight to an imposing light.

ALADDIN 145

The series called 'Aladdin' is based on collections of different coloured glassware found in flea markets, car boot sales and junk shops. The groupings of glassware, although different in shape, texture and function, are united by their respective colour. They are housed in lightbox vitrines resembling museum showcases, which would normally be used to display artefacts but here actually function as floor lights and tables. When illuminated the glassware magically glows like jewels. Although the glassware is presented in a precious way giving the objects a treasure like appearance, 'all that glitters is not gold'.

SPIDER 146

Taking inspiration from spiders and formed by circling of convex rods around a coloured bulb, Spider stands out from the crowd with its sculptural form. When switched on, the coloured bulb emits the light backwards and the lamp casts giant shadows on the surface. A surprising effect that comes as a bonus… Wow!

Page: Descriptions:

148 OCTOPUS CHANDELIER

With aluminum antlers reminding the arms of an octopus, this pendant lamp is an alternative and contemporary version to the traditional chandeliers. Instead of candles you have soft light bulbs. The adjustable arms enable the lamp to be used in varied positions. The sculptural form just adds entertaining spirit to the design. There are two different versions – Single or Double-decker. How much light do you need?

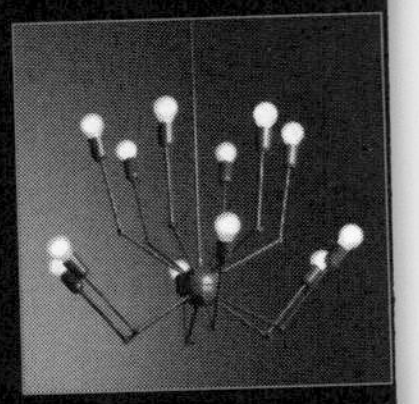

150 MILKII

Milkii, milk jug substitutes the glass of the milk bottle with fine bone china, celebrates the form and cultural associations of the bottle, its relationship with the English breakfast and the taboo of the milk bottle on the breakfast table.

151 PIN LIGHT

This light has a free blown, lead crystal bowl containing steel or glass headed pins. The light bulb is buried in the loose pins and casts a spiky shadow when lit.

152 LORES CHAIR

The digital world often forces tradeoffs between two opposing preferences: speed and quality. When working with images, speed often translates to lower resolution, compressing elements to simplified shapes. During this process, it is the computer, compelled by necessity that makes the decisions about what is taken out and what is left. If applying this bargain to existing products, certain aspects of the product would come to the front, while others recede.

Descriptions: Page:

THE FINAL WHISTLE 153

It is a 1:18 scale, stainless steel whistle chair / sculpture. Symbolizing the end to Alex Garnett's work with scale.

HEY, CHAIR, BE A BOOKSHELF! 154

'Hey Chair Be A Bookshelf!' assembly of products from people´s interiors. Together they become a bookshelf.

LAPJUICER 156

The Lapjuicer is intended to be used by a lap-dancer in a club. The performer uses their body with the object in order to extract fruit juices that can be mixed with alcoholic drinks and consumed by one or more spectators.

Imagine someone making your favourite cocktail with a Lapjuicer. It's for girls and boys – who do you want to see using it?

The Lapjuicer embodies a step beyond the strict 'look but don't touch' policies enforced in many clubs – 'you can't touch - but you can drink my juice…'.

BABY SITTER 158

UNDERDOGMA is a selection of products in which the designers question 11 dogma's in design on 11 products. For this project, they show 9 of the 121 possible outcomes. This results in new products amongst which the Soft Stove of inflammable textiles and a scandalous Baby Sitter. The Baby Sitter is about form and function, the association to it together with the question whether designers have a responsibility in that.

Page: Descriptions:

160 GROW UP — JET COFFEE TABLE

The Plane was designed to respond an exhibition. The theme was based around national identity, in particular with reference of being a practicing designer in Ireland. It is also in response to design education on a national level, in particular, furniture — for many, the extent of design education is literally hawing on a quickly made paper plane prior to firing it off in a boring maths class. The plane is a scaled up traditional A4 paper plane and it is fabricated from 2mm folded, powdercoated mild steel, and is available with optional silicone inset. It is fabricated in France.

162 PET PLUS

When there's no time for a proper meal, vitamins can be bought as health supplements. But what happens when there's no time for relationships? Let us try to imagine when certain hyperreal dog relationships become a commodity product and consumers have the choice to decide which experience to indulge in. Specific relationships can perhaps be purchased as supplements to aid one in fulfilling certain needs, resulting in a new form of hyperreality. Pet plus offers various packages acting as social supplements encouraging further interaction between humans and their dogs.

166 CHA

An enlarged broken mug, functions as a chair or table. It is suitable for use indoor and outdoor.

168 FUNKEY

Computers seemed a perfect contemporary product ripe for subversion. It quite simply dominates our lives and proclaims to empower us with technological excellence, but simultaneously succeeds in making life more confusing, complex and strange than ever. The designer liked the idea of taking something that illustrated man's massive technological advancement and using it in the most primitive manner conceivable, the way primitive man might have used a rock or tree stump. FUNkeys are individually produced to the customer's specific choice of colour and command. It is suitable for use outdoors and are stackable.

Title
"ELECTRIC JOEY" TV STAND
Dimension
50 X 50 X 100 CM
Design
JEREMY WALTON
Client
WALTON.DK
Year
1999

Title
WOODEN USB STICK
Dimension
APPROX. 2 X 2 X 10 CM
Design / Manufacture
OOOMS,
KARIN VAN LIESHOUT
Client
WORLDWIDE
Quantity
1000+
Year
2006

> See also:
Page 090, 101, 134, 152, 234-235

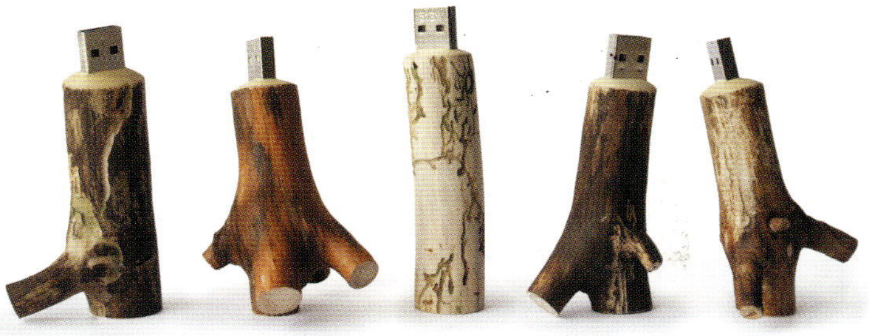

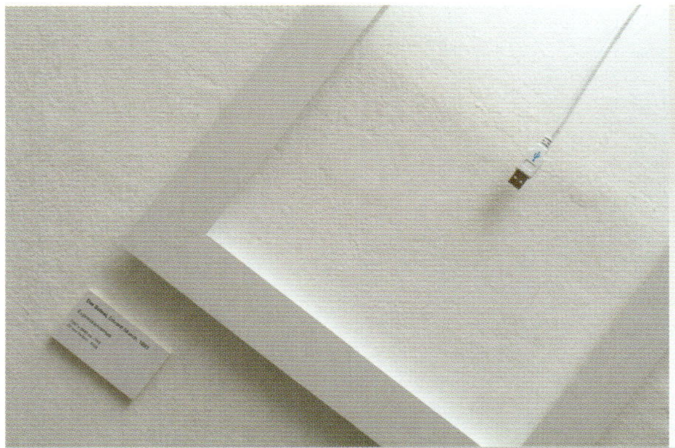
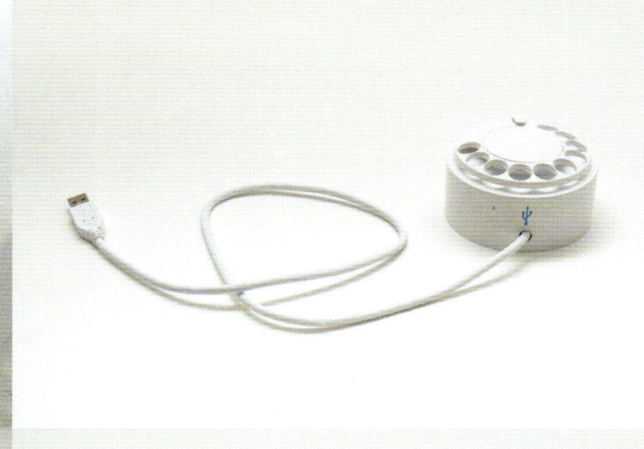

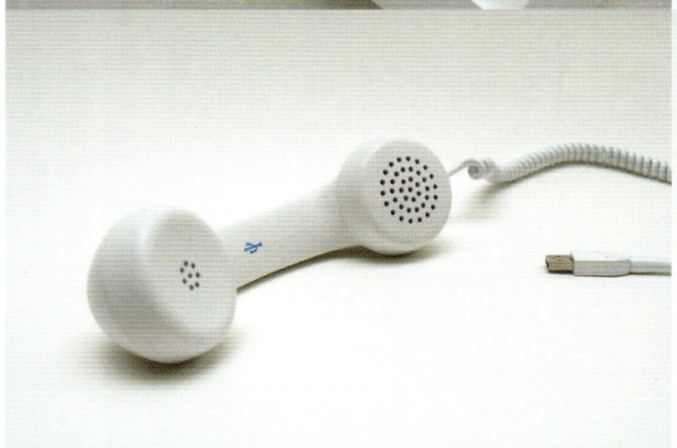
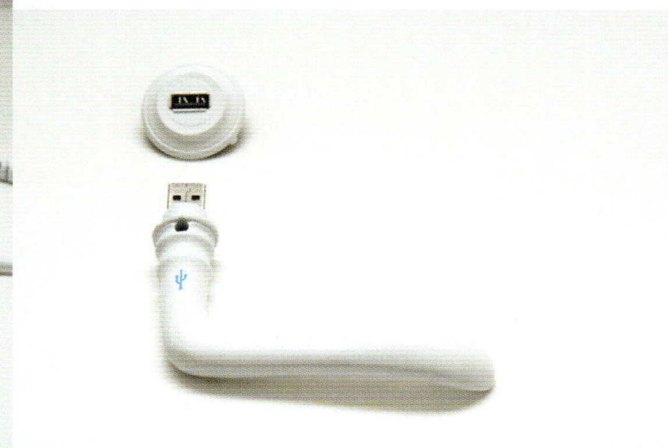

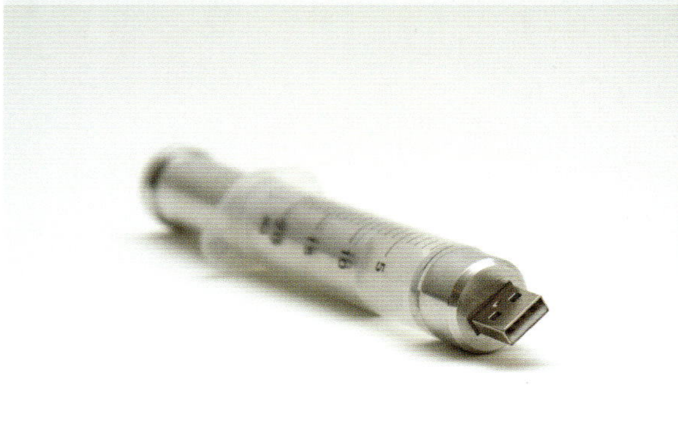
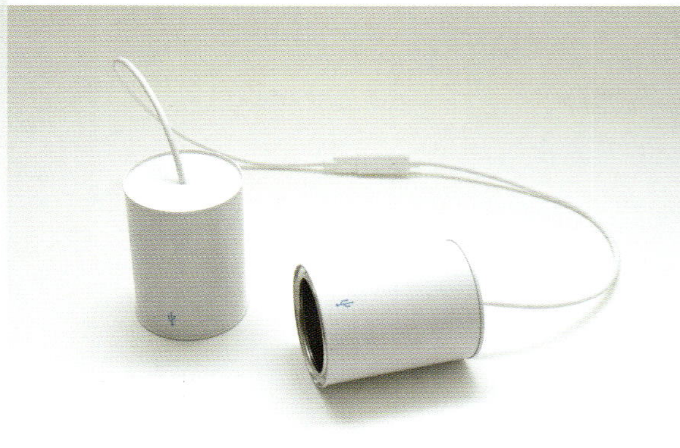

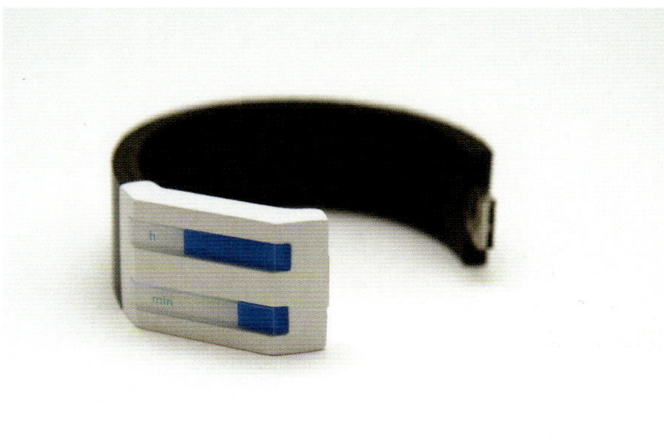
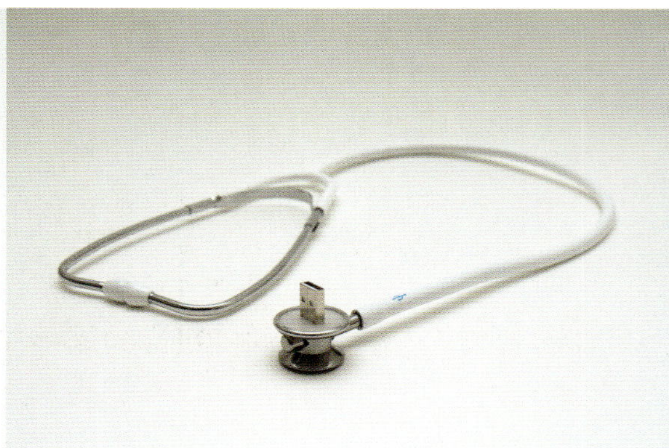

Title

UNIVERSAL CONNECTIONS — OBJECTS BETWEEN ANALOG AND DIGITAL

Dimension

VARIOUS

Design / Client

DIALOG05

Photo

JENS KEUNECKE, ROY HESSING

Quantity

PROTOTYPE

Year

2006

> See also:
Page 092

Title

EARPHONE SPEAKERS 1

MY TURN-ONS 2

Dimension

22.3 CM DIAMETER,
CABLE 1000 CM 1

PLUGS : 5 X 5 X 4.5 CM,
SOCKETS : 9 X 9 X 3 CM,
CRUCIFIX SOCKET :
11.5 X 10 X 3 CM 2

Design / Photo / Client / Manufacture

MIXKO

Quantity

PROTOTYPE

Year

2005 1

2003 2

> See also:
Page 020, 026-027,
095, 153, 166-168

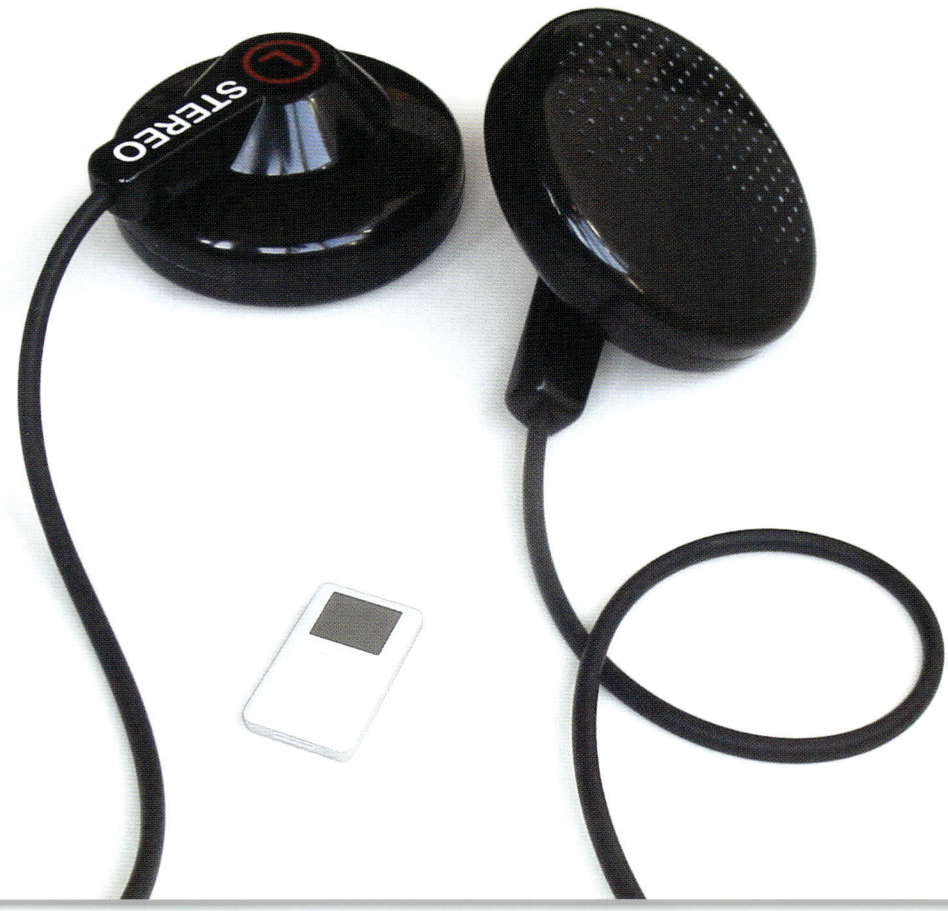

2

1

2

Title
SWAROVSKI —
PENDELUM LIGHT 1
TAPE IT! 2

Dimension
15 X 28 X 12 CM 1
2 X 5.5 CM DIAMETER 2

Design
NICOLAS LE MOIGNE,
AUGUSTIN SCOTT
DE MARTINVILLE,
ADRIEN ROVERO
STUDIO / ECAL 1
AUGUSTIN SCOTT
DE MARTINVILLE,
ADRIEN ROVERO STUDIO
/ ECAL 2

Photo
ANOUS ABRAR

Client
ECAL / SWAROVSKI

Quantity
PROTOTYPE

Year
2005

> See also:
Page 042-043, 045, 088,
096, 102, 238-239

1

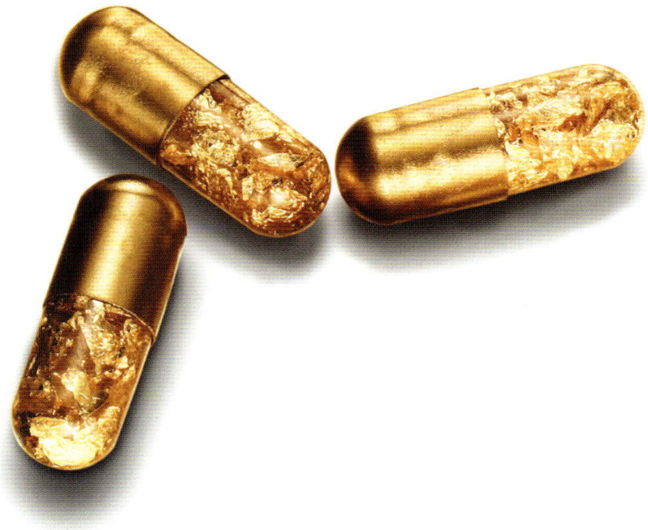

2

3

Title

18 KT GOLD BIC PEN LID 1
GOLD PILLS 2
ULTIMATESETTING
AKA KILLERRING 3

Dimension

6.4 CM 1
1.9 CM 2

Design / Photo / Client

TOBIAS WONG

Year

2005 1,2
2004 3

> See also:
Page 099

Title
SMOKINGMITTENS
Dimension
25.4 X 15.2 CM
Design / Photo / Client
TOBIAS WONG
Quantity
500
Year
2002

> See also:
Page 098

Title

LAST BUNCH OF FLOWERS FOR MY GIRL-FRIEND (CAL .357 MAGNUM)

Dimension

CAL.357 MAGNUM

Design / Client / Manufacture

MICHAEL SANS BERLIN

Quantity

SET OF 5, LIMITED EDITIONS (250)

Year

2005

> See also:
Page 124-125

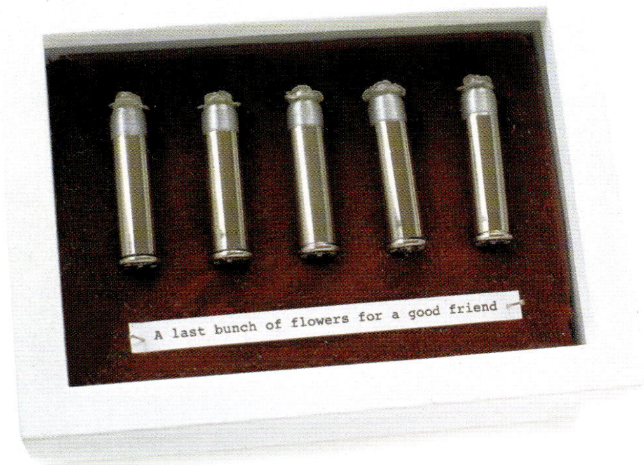

Title
DUTCH DELIGHT
Dimension
APPROX. 4 X 4 X 23 CM
Design
OOOMS,
DAVY GROSEMANS
Client
WORLDWIDE
Manufacture
ANOUK OMLO
Quantity
100+
Year
2002

> See also:
Page 090-091, 134,
152, 234-235

Title
DIP 1/3
Dimension
VARIOUS
Design
ADRIEN ROVERO STUDIO / ECAL
Photo
MILO KELLER
Client / Manufacture
CHRISTOFLE
Year
2006

> See also:
Page 042-043, 045,
088, 096-097, 238-239

1

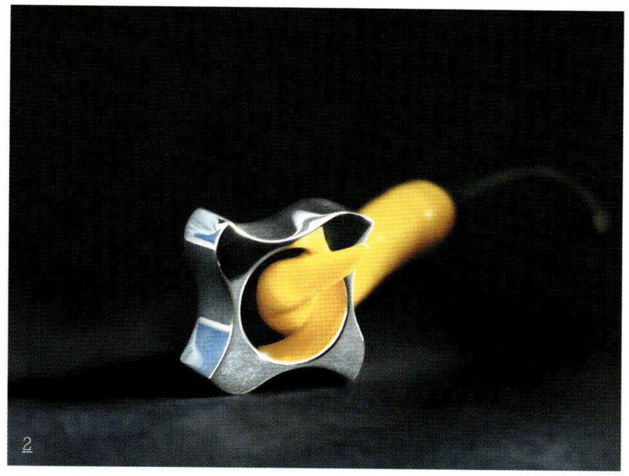
2

3

4

Title
THE SUN,
THE EARTH,
THE MOON 1
THE KISS 2

Dimension
APPROX. 3.4 X 3.4 X 1.4 CM 1
APPROX. 3.8 X 3.2 X 1.4 CM 2
(DEPENDING ON FINGERSIZE)

Design / Manufacture
GLITNER

Photo
PATRICIA RODAS

Quantity
PRODUCE ON REQUEST

Year
2003

Title
TANK YOU
Dimension
16 X 10 X 11 CM
Design
LA TÊTE AU CUBE
Quantity
800
Year
2006

> See also:
Page 105

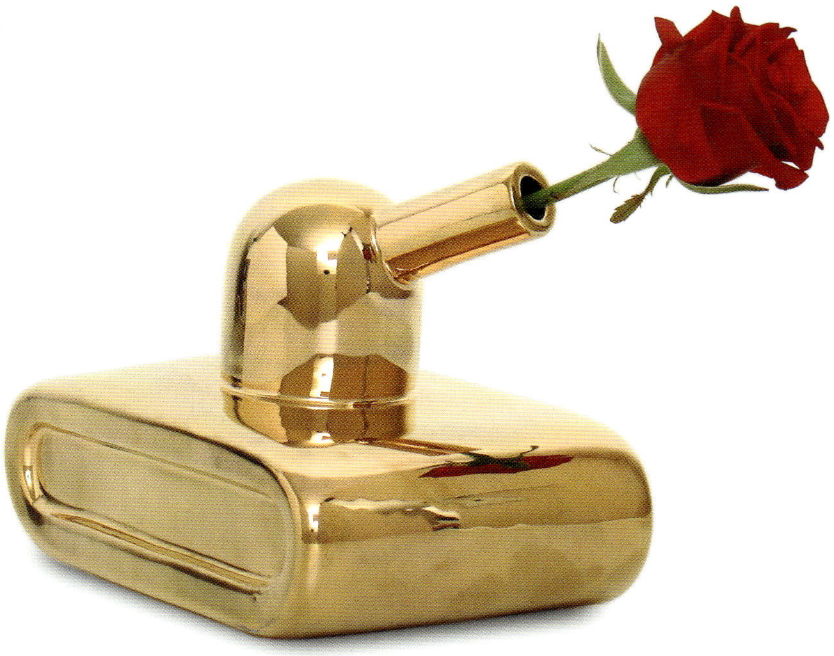

Title
PROAESTHETICS
— EYE PATCH
Dimension
7 X 4 CM
Design / Client / Manufacture
DAMIAN O'SULLIVAN DESIGN
Photo
ADRIAAN VAN DER PLOEG
Quantity
3
Collection
EXHIBITED AT
THE BOIJMANS VAN
BEUNINGEN MUSEUM
IN ROTTERDAM
Year
2000

> See also:
Page 107-109

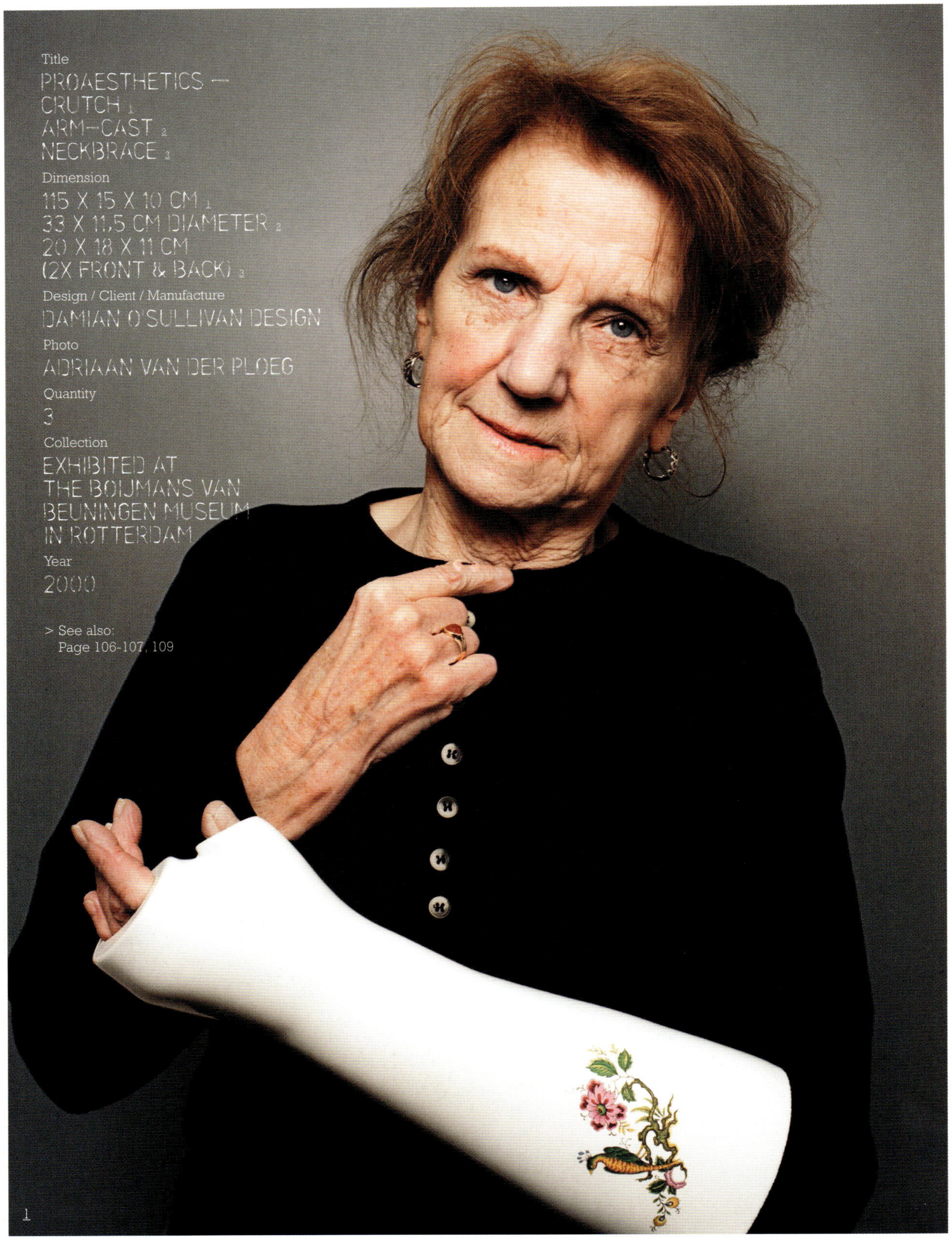

Title
PROAESTHETICS —
CRUTCH 1
ARM-CAST 2
NECKBRACE 3

Dimension
115 X 15 X 10 CM 1
33 X 11,5 CM DIAMETER 2
20 X 18 X 11 CM
(2X FRONT & BACK) 3

Design / Client / Manufacture
DAMIAN O'SULLIVAN DESIGN

Photo
ADRIAAN VAN DER PLOEG

Quantity
3

Collection
EXHIBITED AT
THE BOIJMANS VAN
BEUNINGEN MUSEUM
IN ROTTERDAM

Year
2000

> See also:
Page 106-107, 109

1

2

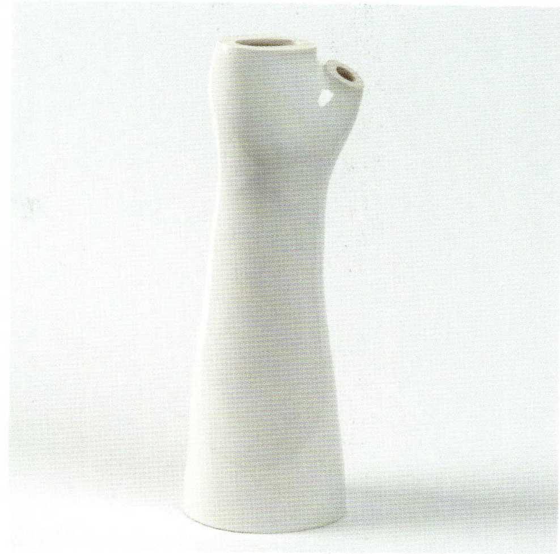

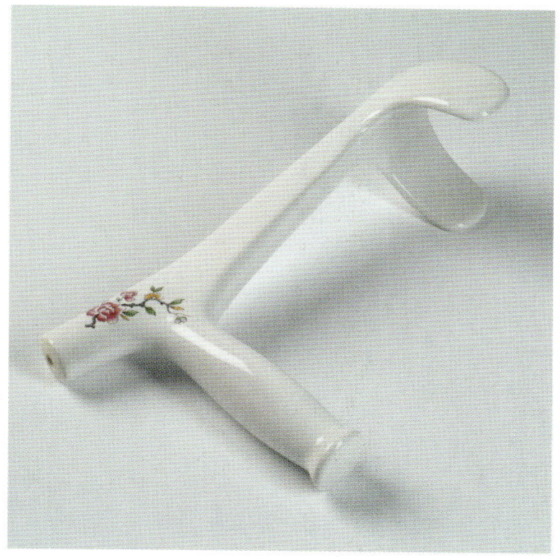

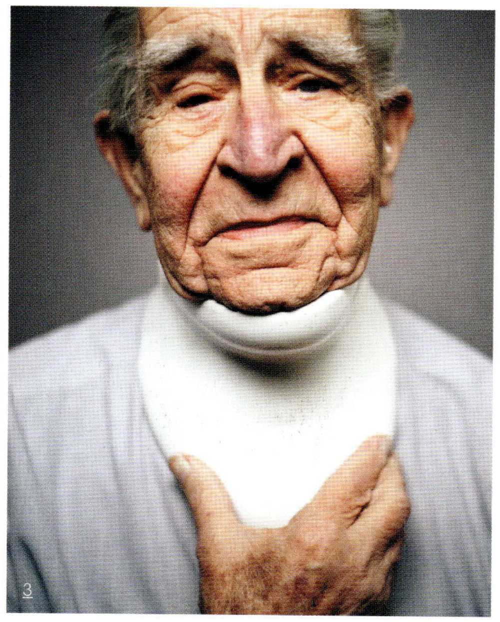

3

Title

HEAD HAND BAGS

Dimension

HUMAN HEAD SIZE

Design / Manufacture

YAEL MER

Photo

MATHIAS HAHN

Year

2006

> See also:
Page 110, 208-209

Title
INTERNAL WATCH
Dimension
23.5 X 5 X 1.5 CM
Design
VLAEMSCH ( )
Client
LICENSED BVBA
Year
2006

> See also:
Page 024-025, 050-051, 119

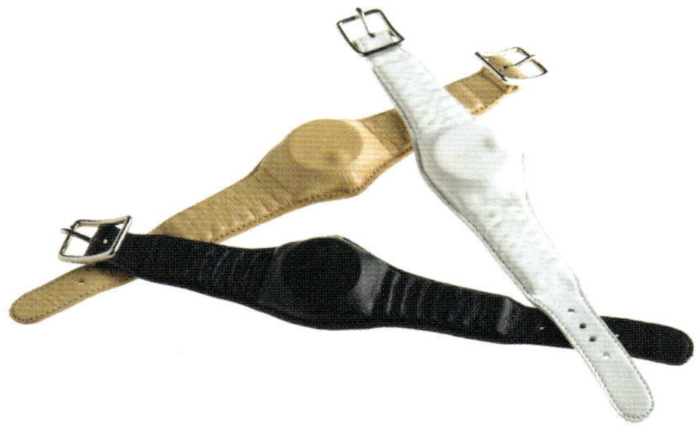

Title

CHARACTERIZED DESIGN — ICON WATCH

Dimension

3.45 X 22 X 0.85 CM

Design

&DESIGN

Manufacture

IDEA INTERNATIONAL

Year

2007

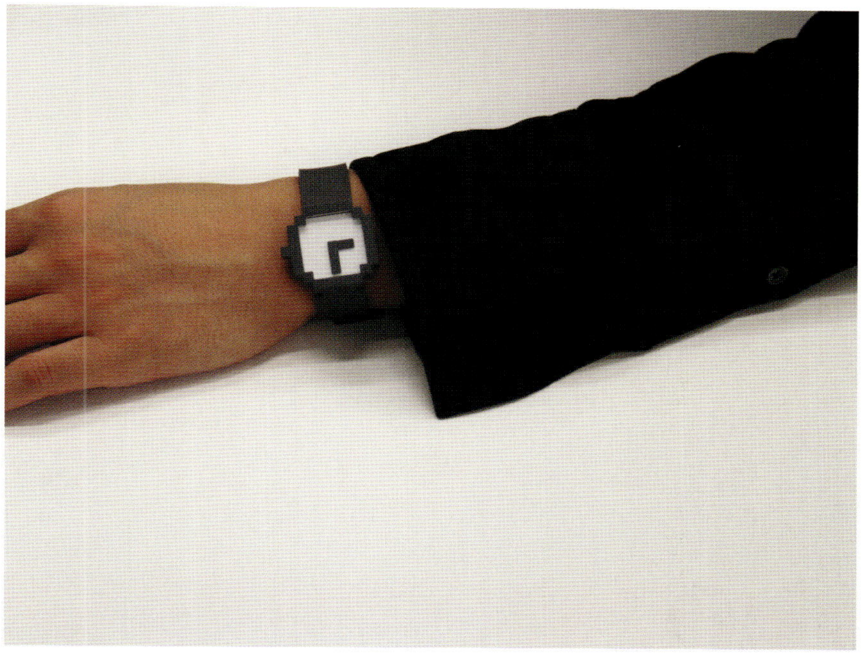

1

2

Title

PEEPSHOW 1 GUARDIAN ANGEL 2
NOT-A-LAPTOP 3 HUNGRY JACK 4

Dimension

30 X 40 X 10 CM 1 33 X 26 X 8 CM /
43 X 31 X 10 CM 2 37 X 47 X 10 CM 3
39 X 23 X 5CM 4

Design / Photo / Manufacture

VLIEGER & VANDAM

Quantity

100 1 2000 2 75 3 150 4

Collection

PERMANENT COLLECTION OF
THE MUSEUM OF MODERN ART,
NEW YORK 2

Year

2005 1,4 2004 2 2006 3

> See also:
Page 114

4

3

Title
LADY'S BAG ME & MY BERETTA
Dimension
20 X 20 X 8 CM
Design / Client
STUDIO VAN EIJK &
VAN DER LUBBE
Photo
STUDIO 4/A,
PEER VAN DE KRUIS
Year
1999

> See also:
Page 070-071, 117, 158-159

Title

HIDE & SHOW

Dimension

170 X 90 X 52 CM,
170 X 60 X 45 CM

Design / Client

JUDITH SENG

Quantity

PROTOTYPE

Year

2004

Title
BOOTBAG
Dimension
9 X 24 X 26 CM
Design
VLAEMSCH ( )
Client
LICENSED BVBA
Year
2004

> **See also:**
Page 024-025, 050-051, 112

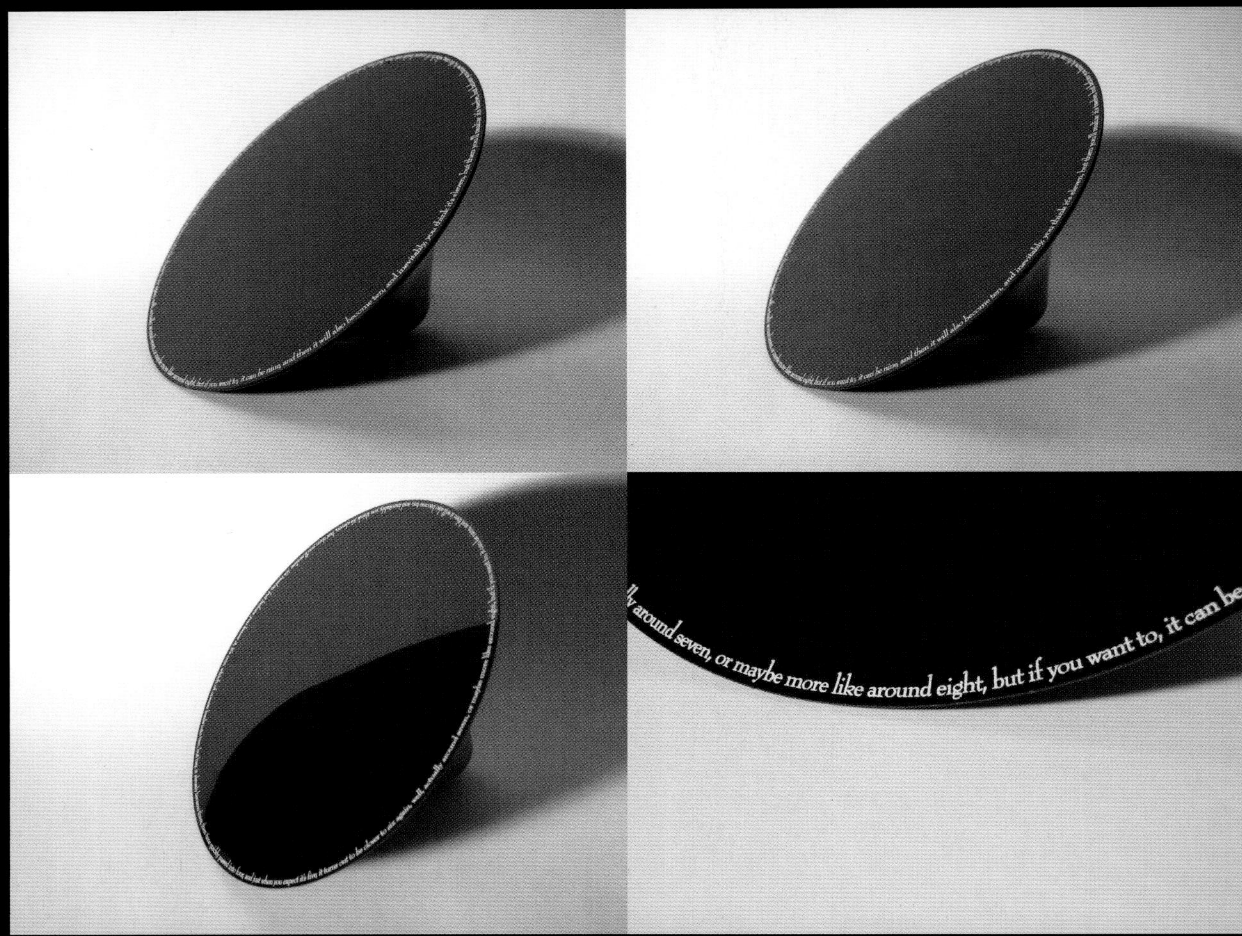

Title
ABOUT TIME
Dimension
10 X 25 CM
Design
BURO VORMKRIJGERS
Photo
SANDER MULDER
Manufacture / Collection
CULTIVATE
Year
2006

> See also:
Page 044, 122-123, 135,
194-195, 218-219

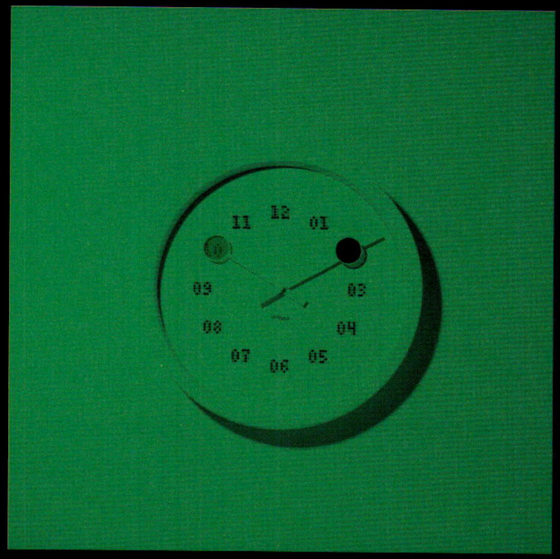

Title
SEEING DOUBLES
Dimension
40.6 CM
Design / Manufacture
SEEDZ LTD.
Photo
TIM LAU
Client
CHILLICHILLY
Awards
SILVER,
HKDA AWARD 05
Year
2005

Title
PONG CLOCK
Dimension
18 X 12 X 3 CM
Design / Manufacture
BURO VORMKRIJGERS
Photo
JASPER DE LEEUW
Client
IN LICENSE OF ATARI INC.
Quantity
400
Year
2006

> See also:
Page 044, 120, 123,
135, 194-195, 218-219

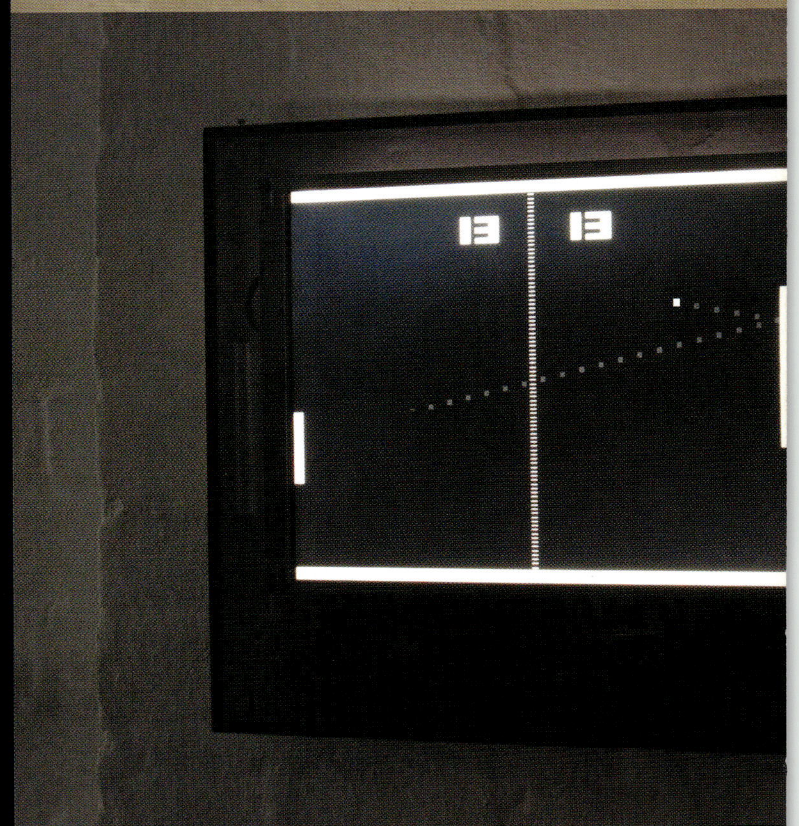

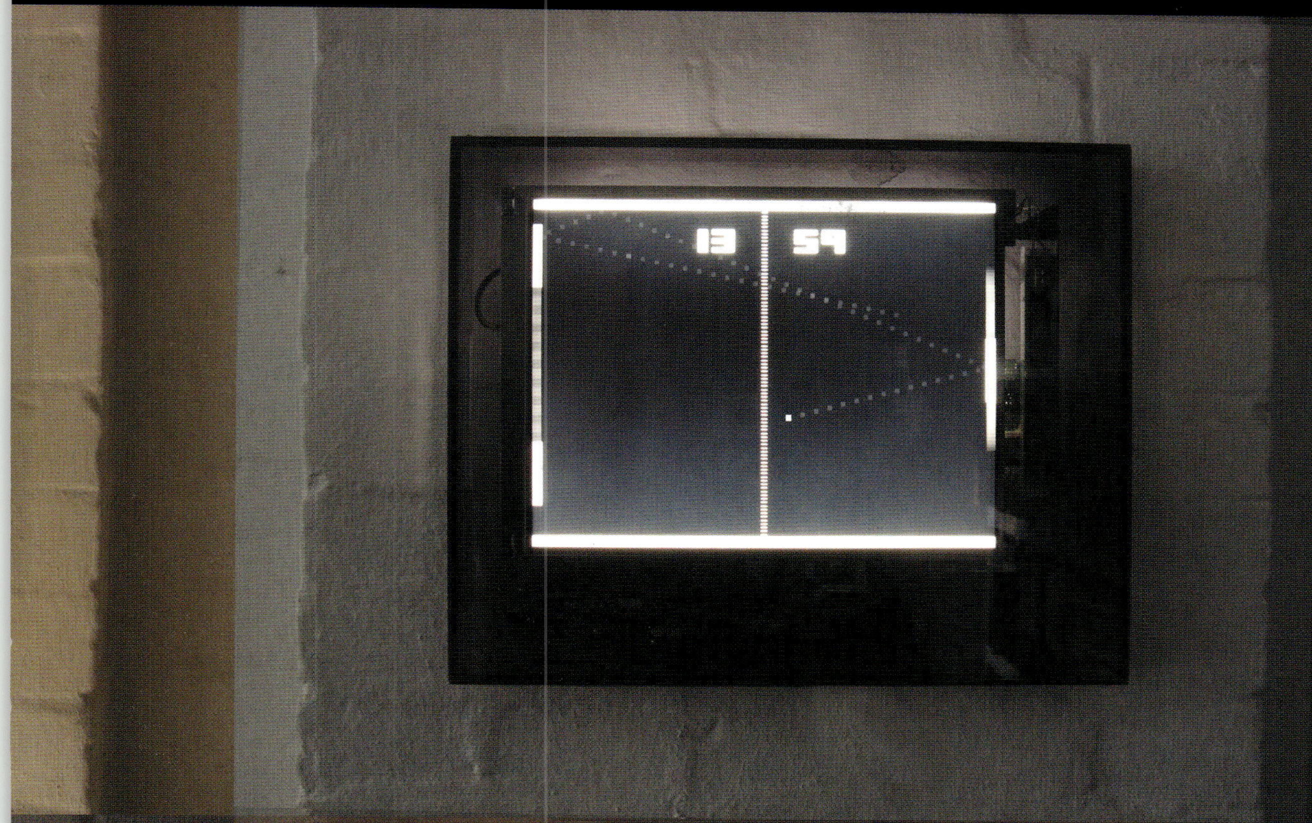
13 59

14 12

Title
CUCKOU
Dimension
54 X 36 X 20 CM
Design / Client / Manufacture
MICHAEL SANS BERLIN
Photo
DOMINIK BUTZMANN
Quantity
PROTOTYPE
Year
2006

> See also:
Page 100, 125

Title

FLYING FISH BOWL

Dimension

60 X 60 X 30 CM

Design / Client

SHAY ALKALAY

Quantity

PROTOTYPE

Awards

CONCEPT PRODUCT FINALIST,
IDEAL HOME SHOW,
LONDON 2007

Year

2006

> See also:
Page 127, 174, 176-177

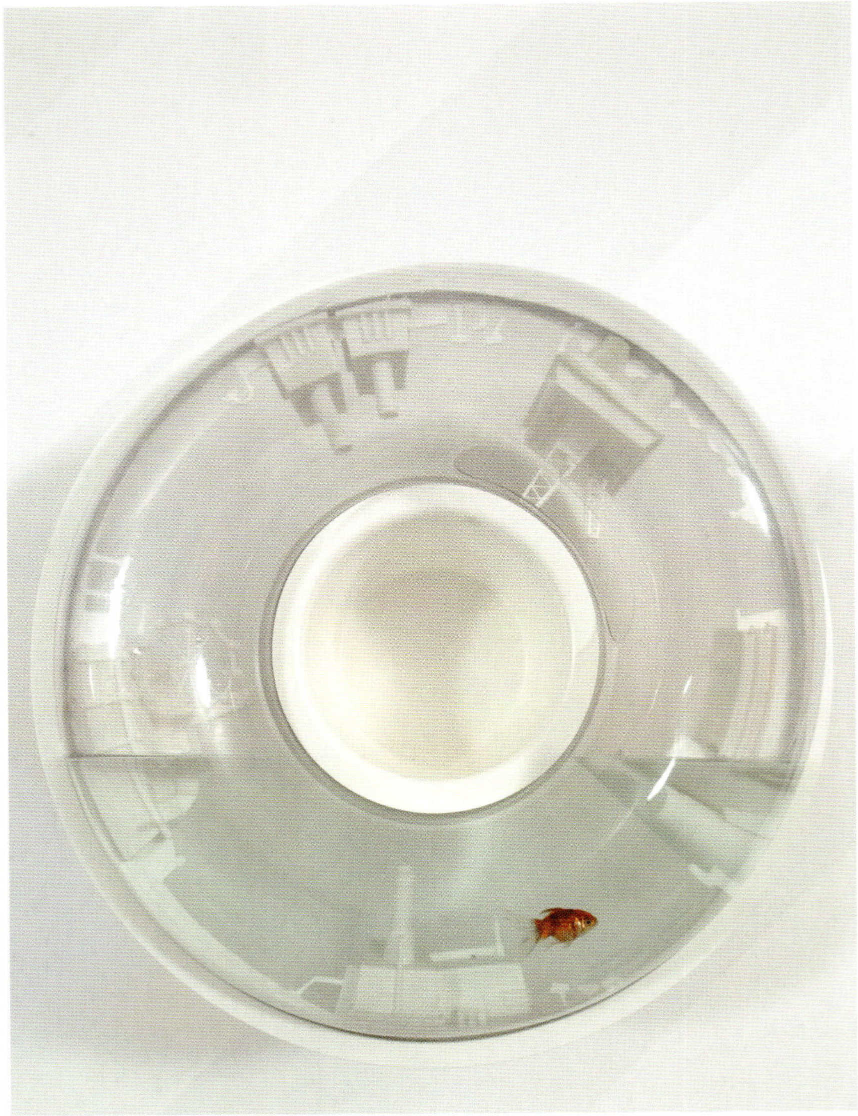

Title
DIP
Dimension
32 X 29 X 5 CM
Design / Photo / Client / Manufacture
STAUFFACHERBENZ
Quantity
20
Year
2004

> See also:
Page 214-215

Title
16 ANIMALI COMPONIBILI
Dimension
38 X 27 X 5.5 CM
Design
DANESE S.R.L.
Photo
GIO PINI, MIRO ZAGNOLI
Quantity
LIMITED EDITIONS

Title
WAR BOWL SERIES
Dimension
13 X 44 CM DIAMETER
Design
DOMINIC WILCOX
Manufacture
THORSTEN VAN ELTEN
Year
2002

> See also:
Page 131

Title
NAPKING
Dimension
47.5 X 47 CM
Design / Client
BLEACH DESIGN, LDA
Photo
SERGIO CLARO
Quantity
5000
Collection
GENERAL CATALOGUE
Year
2005

Title

VOODOO
KNIFE BLOCK

Dimension

42 X 36 X 26 CM

Design

RAFZ DESIGN STUDIO

Manufacture

VICEVERSA

Quantity

APPROX. 1,000,000
IN 2 YRS

Year

2003

Title
GLASSBULBS
Dimension
APPROX. 5 X 5 X 25 CM
Design / Manufacture
OOOMS
Quantity
PROTOTYPE
Year
2003

> See also:
Page 090-091, 101,
152, 234-235

Title
CECI
Dimension
11 X 6 CM
Design / Manufacture
BURO VORMKRIJGERS
Photo
SANDER MULDER
Quantity
250
Year
2005

> See also:
Page 044, 120, 122-123,
194-195, 218-219

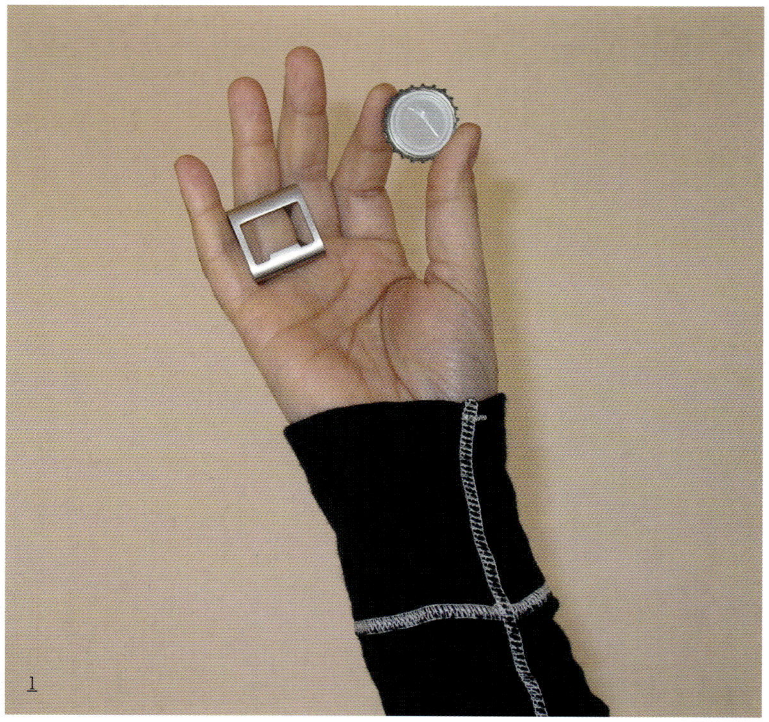

1

2

Title
BAR MAN RING 1
POCKET BRUSH 2
FLAMP 3

Dimension
3 X 3 X 3 CM 1
8 X 3.5 CM 2
5 X 15 CM 3

Design
CONCEPTS AND IDEAS FOR COMMERCIAL PURPOSES S.L.

Photo
IMAGEKONTAINER / INGA KNÖLKE 1.3

Client / Manufacture
THE ORIGINAL CHA-CHÁ, BARCELONA 1
DIM (DIE IMAGINÄRE MANUFAKTUR), BERLIN 2
GALERIA H2O, BARCELONA 3

Quantity
MASS PRODUCTION 1.3

Year
2003 1
2000 2
1998 3

> See also:
Page 137, 186

3

Title
ANIMALS*
Dimension
VARIOUS
Design
FRONT
Photo
MAARTEN VAN HOUTEN
Manufacture / Collection
MOOOI
Quantity
HORSE IS IN PRODUCTION; PIG AND RABBIT ARE STILL IN DEVELOPMENT
Year
2006

* Provided by Moooi

Title

KEBAB LAMP

Dimension

APPROX. 170 CM
(HEIGHT)

Design / Manufacture

COMMITTEE

Quantity

APPROX. 15
PER YEAR

Collection

VARIOUS PRIVATE

Awards

ELLE DECORATION UK
LIGHTING AWARD 2004

Year

2004

Title
DISPOSABLE CHANDELIER
Dimension
200 X 50 CM DIAMETER
Design / Manufacture
STUART HAYGARTH
Quantity
20
Year
2005

> See also:
Page 141-145

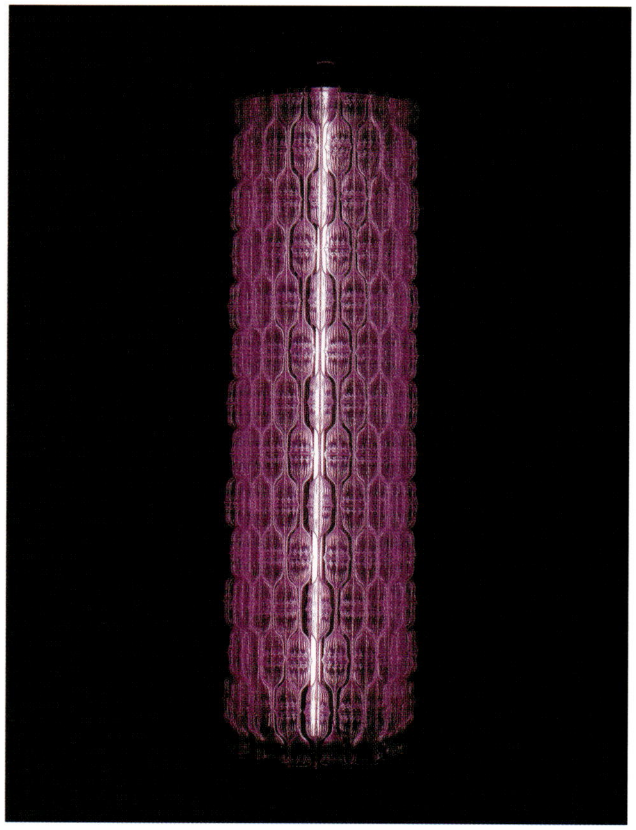

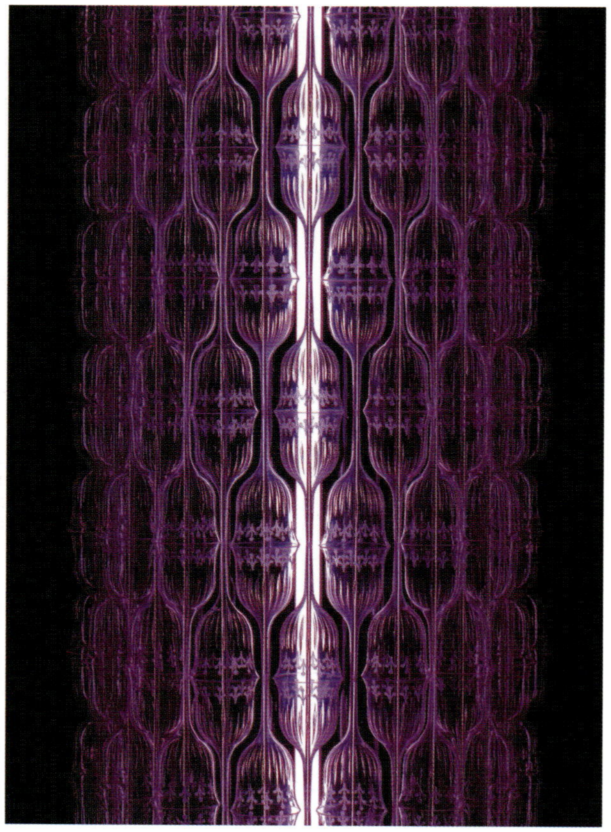

Title
MILLENNIUM CHANDELIER
Dimension
160 X 80 CM DIAMETER
Design
STUART HAYGARTH
Quantity
20
Awards
SHORTLISTED DESIGNBOOM
COMPETITON 2005
Year
2004

> See also:
Page 140, 142-145

Title
TIDE CHANDELIER
Dimension
190–210 X 150 CM DIAMETER
Design / Manufacture
STUART HAYGARTH
Quantity
10
Year
2005

> See also:
Page 140-141, 143-145

Title
TWENTYTWENTY
Dimension
64 X 64 X 56 CM,
46.5 X 46.5 X 45 CM
Design
STUART HAYGARTH
Client
LONDON DESIGN MUSEUM
Quantity
1 LARGE, 2 SMALL
Collection
DESIGN MUSEUM
Year
2006

> See also:
Page 140-143, 145

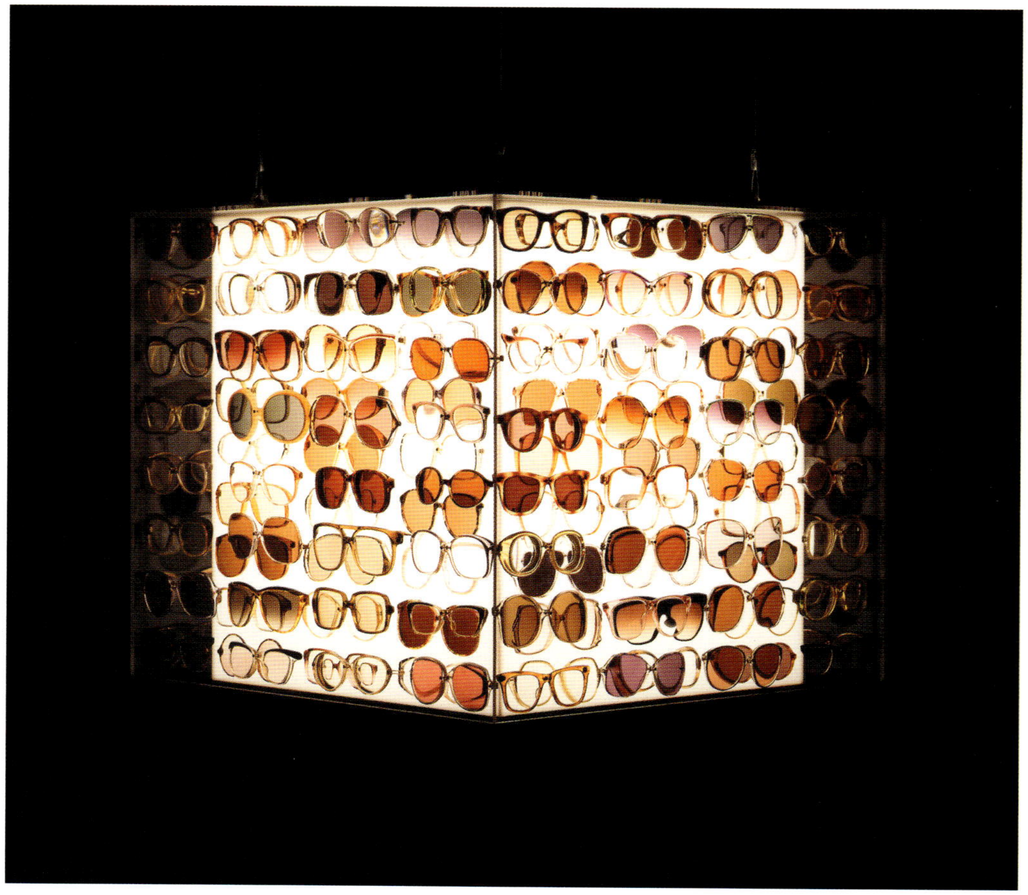

Title
ALADDIN
Dimension
128.2 X 128.2 CM, 97.9 X 97.9 CM,
80.2 X 80.2 CM, 67.4 X 67.4 CM
(WITH AVERAGE HEIGHT 45 CM)
Design
STUART HAYGARTH
Quantity
12
Year
2006

> See also:
Page 140-144

Title
SPIDER
Dimension
40 X 60 X 80 CM
Design / Manufacture
AUTOBAN
Photo
TAMER YILMAZ
Year
2003

> See also:
Page 147-149

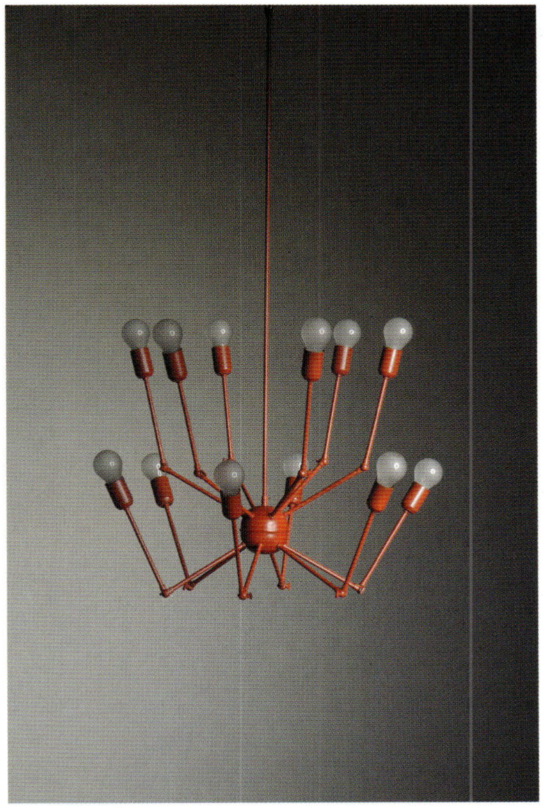

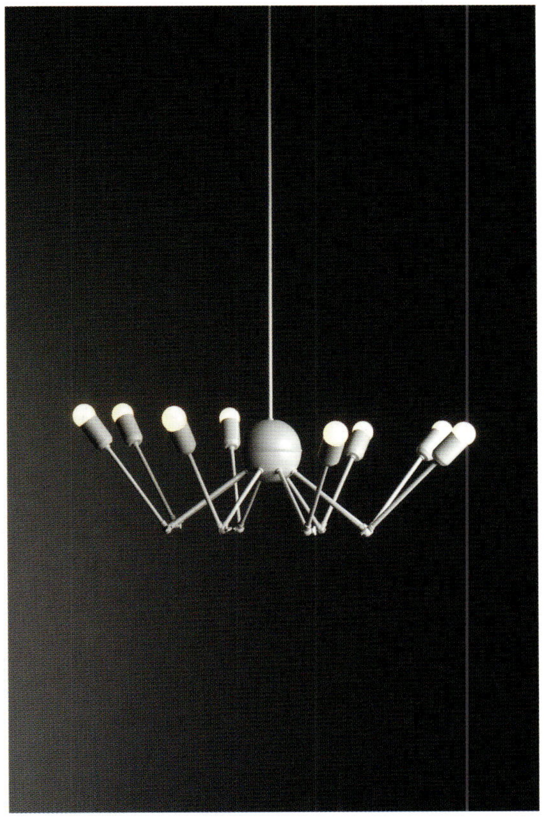

Title
OCTOPUS CHANDELIER
Dimension
80 CM RADIUS
Design / Manufacture
AUTOBAN
Photo
TAMER YILMAZ
Year
2005

> See also:
Page 146-148

Title
MILKII
Dimension
14 X 8 CM
DIAMETER
Design / Client / Manufacture
MULTI
Photo
LEE FUNNELL,
GRAPHICPHOTO
Quantity
500+
Collection
PRIVATE COLLECTIONS
Year
2002

> See also:
Page 151

Title
PIN LIGHT
Dimension
14.5 X 14 CM DIAMETER
Design / Client / Manufacture
MULTI
Photo
LEE FUNNELL, GRAPHICPHOTO
Quantity
UNDER 10
Year
2002

> See also:
Page 150

Title
LORES CHAIR
Dimension
80 X 80 X 60 CM
Design / Manufacture
OOOMS
Quantity
PROTOTYPE
Year
2000

> See also:
Page 090-091,
101, 134, 234-235

Title
THE FINAL WHISTLE
Dimension
87 X 45 X 57 CM
Design / Photo / Manufacture
MIXKO
Client / Collection
PAUL SMITH
Quantity
10
Year
2005

> See also:
Page 020, 026-027,
094-095, 166-168

Title
"HEY, CHAIR,
BE A BOOKSHELF!"

Dimension
210 X 115 X 90 CM

Design
MAARTEN BAAS

Photo
M. VAN HOUTEN

Manufacture
BAAS AND DEN HERDER

Quantity
EVERY PIECE
IS A UNIQUE PIECE

Year
2005

> See also:
Page 155

Title
LAPJUICER
Dimension
100 X 50 CM
Design / Photo / Client / Manufacture
THEO HUMPHRIES (3EYES), PHILIP WORTHINGTON
Quantity
2
Awards
EXHIBITED DURING "TOUCH ME" AT THE VICTORIA & ALBERT MUSEUM, LONDON UK, SUMMER 2005
Year
2004

> See also:
Page 157

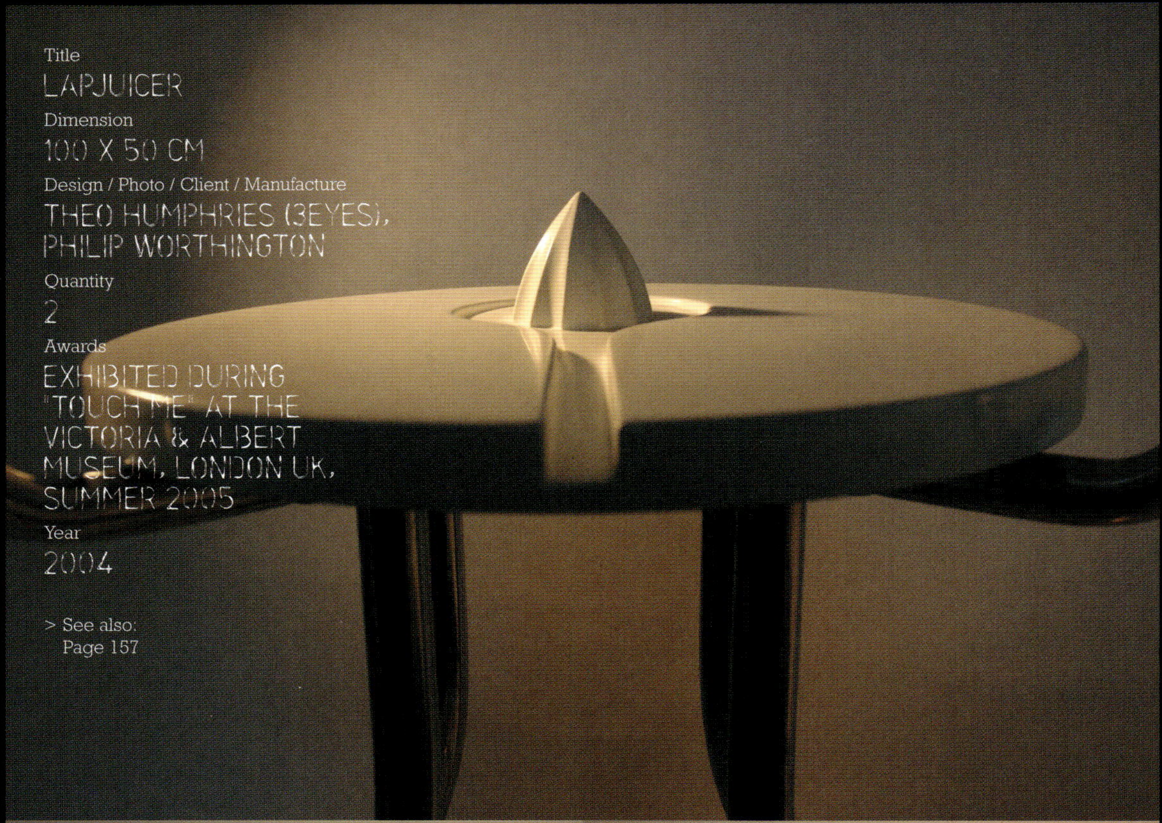

Title
BABY SITTER
Dimension
40 X 40 X 120 CM
Design / Client
STUDIO VAN EIJK & VAN DER LUBBE
Photo
ELI (POODLE MILAAN)
Collection
UNDERDOGMA
Year
2004

> See also:
Page 070-071, 116-117, 159

INTERNI

INTERNI

Title
GROW UP —
JET COFFEE TABLE
Dimension
2000 X 900 X 320 CM
Design / Client
FIFTY-EIGHT B
Photo
CECIL MATHIEU
Manufacture
TPRP, FRANCE
Quantity
50
Year
2006

> See also:
Page 046-047, 160

Title
PET PLUS
Design
ALICE WANG
Year
2006

> See also:
Page 162, 164-165

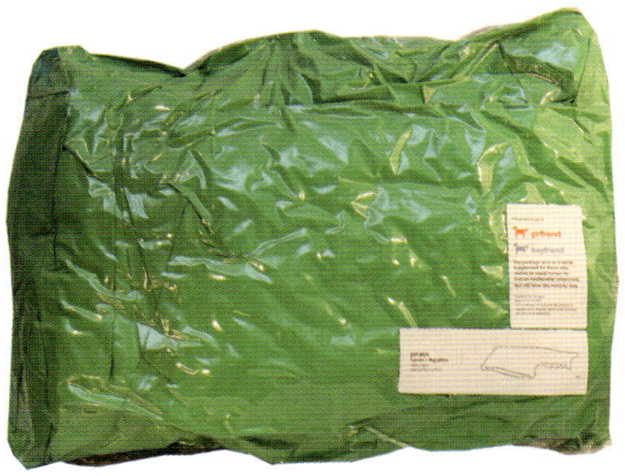

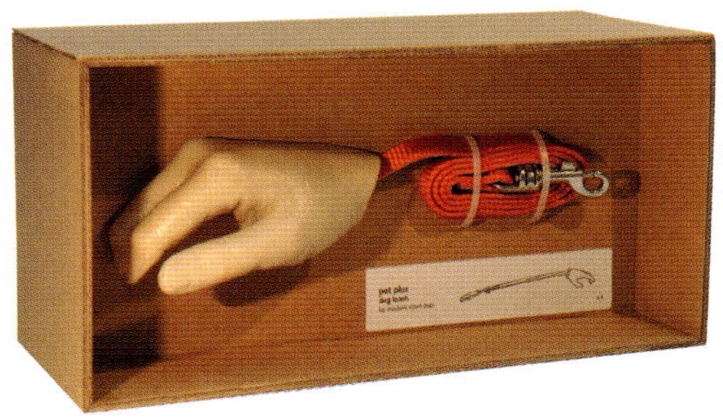

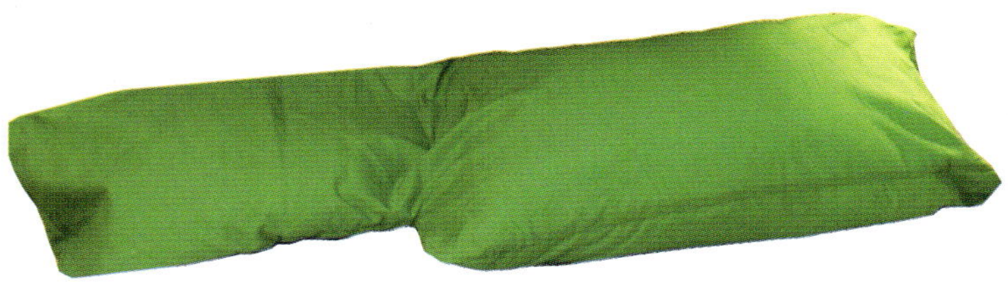

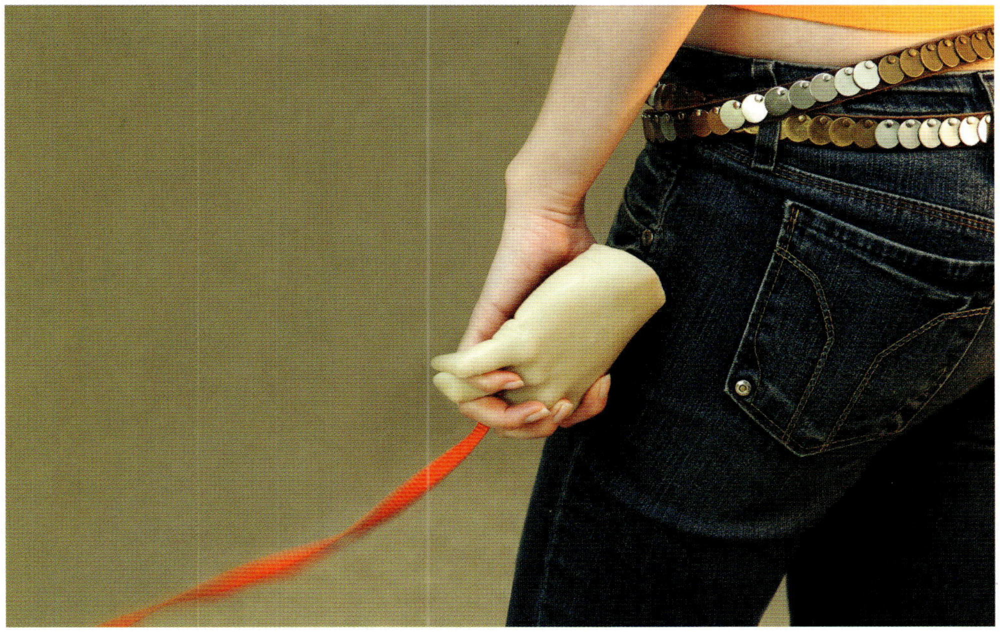

Title
CHA
Dimension
70 X 82 X 70 CM DIAMETER
Design / Photo / Manufacture
MIXKO
Client
TOPSHOP
Quantity
PRODUCE ON REQUEST,
"I LOVE NY" EDITION : 10
Year
2004

> See also:
Page 020, 026-027,
094-095, 153, 167-168

Title
FUNKEY
Dimension
40 X 58 X 58 CM DIAMETER
Design / Photo / Client / Manufacture
MIXKO
Quantity
100+ (PRODUCE ON REQUEST)
Awards
THE BRITISH COUNCIL,
"RISING LIGHTS" 2003,
THE GERMAN DESIGN COUNCIL,
"INSPIRED BY COLOGNE" 2004
Year
1999

> See also:
Page 020, 026-027,
094-095, 153, 166-167

INTERACTIVE + SENSATION

TECHNOLOGY PERVADES OUR LIVES, WHETHER WE WANT IT OR NOT. TO MAKE ALL THIS NEW TECHNOLOGY APPROACHABLE AND ENJOYABLE, AS WELL AS USEFUL BY HUMANS IT NEEDS TO BE DESIGNED HAVING IN MIND THE IMPLICATIONS IN HUMANS. DESIGNERS AND USERS ARE, ABOVE ALL, COMPLEX INDIVIDUALS MOVING THROUGH AN EQUALLY COMPLEX ENVIRONMENT. BEYOND AESTHETICS, IT IS ABOUT UNDERSTANDING HUMAN BEHAVIOUR. INTERACTION DESIGN IS ABOUT ENABLING CONNECTIONS BETWEEN PEOPLE, HELPING PEOPLE WITH EFFICIENT WAYS TO COMMUNICATE BETTER BETWEEN THEMSELVES. THAT IS "THE REASON WHY" OF INTERACTION DESIGN. THE OBJECTS PRESENTED ARE EXAMPLES OF EFFECTIVE RESEARCH OF THINKING ABOUT THE LIFE AROUND US.

Page: Descriptions:

169 AFTER DARK

Afroditi's Ring is the rediscovery of the ancient signet ring. A rubber stamp is set on the ring just like a precious stone, secretly leaves imprints when shaking one's hand, touching, on paper or generally, wherever you lay your hands on. It explores gestures, the concept of familiarity as well as the secret coding of handshakes. This piece of jewellery acts like a catalyst of communication between people.

170 SOLITAIRE DISH, O'S AND X'S ASHTRAY, BATTLESHIPS NAPKINS

This set of three objects all combine play with known table-top accessories.

Solitaire Dish - A snack Dish that encourages you to play with your food. / O's and X's Ashtray - Play while smoking. Best played with a smoking partner. / Battleships Napkins - It needs 2 players with a napkin each to play.

(See instructions for how to play for the above all.)

172 'DO DESTROY' GIFT CARDS

It is a series of cards. Gifts are hidden within the chemistry of the paper. You have to destroy the card to reveal the gift, a truly satisfying experience - much like setting fire to an ex-lover's note. The series includes cards that you have to burn, soak in water, blow into and plant.

174 STICKYSTAINS

Iron-On stickers that turn unsightly stains into attractive doodles. An alternative way to prolong the life of stained clothes.

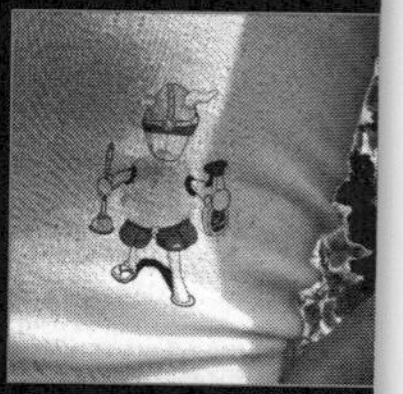

Descriptions: Page:

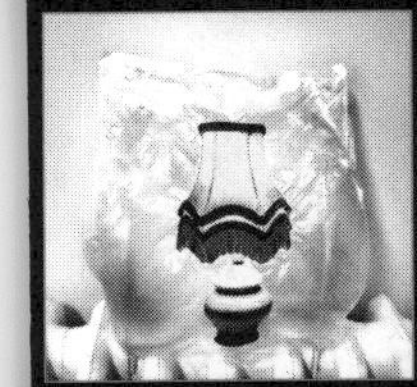

GAS BAGS 175

Radiators are neglected space. Fireplaces have precisely the same function yet we embellish fireplaces with mantles and ornamentation. The bags attach to white powder-coated frames and inflate with the rising warm air, adorning the negative space above the radiator.

HOOVER BAG BEARS 176

This product is a continuation of the Bin Bag Bears project, and keeps the idea of turning waste into a lovely bear. Instead of throwing away your Hoover bag, you have the chance to put a sticker on the Hoover bag lid and then to keep yourself the own personal dust or maybe give it to someone else.

BIN BEAR BAG 177

The objects are part of a myth story that represents the designer's idea of having a bunch of bin BEARS in the streets rather than a bunch of rubbish bags waiting to be collected by the council rubbish disposal services. The idea came up after watching the huge amount of bin bags around London's streets. He thought way not to use this content and turn this daily image into a surreal one something more surprising. Where ever we go we would see bunches of bin bag bears sitting on the pavement next to each other, see how by sitting next to empty bottles and other stuff this situation getting a new context.

PRIVACY SCARF 178

A series of objects that call attention to the privacy we sacrifice when using mobile technological devices in public spaces. They have been designed as visual tools to prompt discussions around issues of privacy, isolation and the implications of handling personal information in public.

Page: Descriptions:

180 PUNCH'N'CUDDLE BAG

The Punch'n'Cuddle Bag is what you would get if you crossed a beanbag with a punching bag. It demands physical and expressive interaction. Whether your emotions are positive or negative, this playful object will absorb them and sometimes return for more! It is the ultimate domestic companion for both lovin' and hatin'.

Caution: this furniture item will fight back!

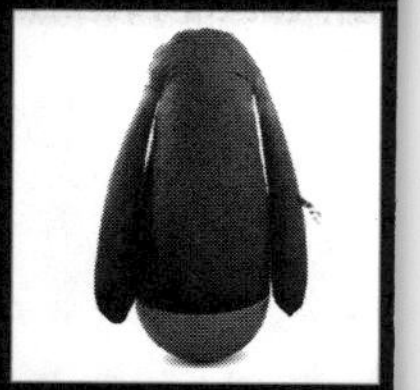

182 PLANT TAPE

A roll of tape is cut so that branches appear. The empty space between the branches become leafs. It allows you to create an artificial plant that needs not watering. It is ever extendable and will take over your wall if you want.

184 GREEN LETTUCE

How come fake plants don't bloom? Bloom! It is a series of fake-plants which illustrate the fascinating growing process of nature. Spray some water onto the plant and the leaves will open at random.

Material: Polyurethane foam

185 CAPILLARY TREE

How come fake plants don't bloom? Bloom! is a series of fake-plants which illustrate the fascinating growing process of nature. Poor some water in the pot and the leaves will open within seven days.

Material: Polyurethane foam

Descriptions: Page:

PLANT-ME PET 186

The Plant-me Pet, with vegetable plant seeds for eyes, is useless until you make it disappear, burying it so that it might grow into an edible food plant. It is a pet that forces you to decide between emotion and function.

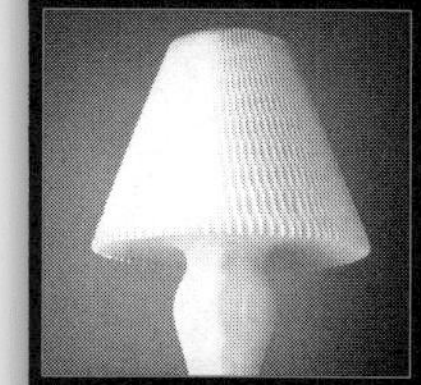

HONEYCOMB LAMP 187

'Denguri-paper' is used for the shade of this lamp. The paper is a special product of Shikoku region in Japan. It was born from the workman's delicate and time consuming work. When it is packed, the thickness is about 2cm. And it can be a lamp shade when you open it and put together those edges by some pins (Pins are attached). It uses a 10W bulb and cord (not included).

JELLYFRIENDS 188

JellyFriends is a quietly pulsating light armature that interacts with the user and reacts when someone comes near. The jellyfish shaped body is moulded in silicone and comes with adjustable silicone cord legs that enable an unique expression of each JellyFriend. JellyFriends explore the boundries between artefact and man and advocates a living, dynamic environment.

ON OFF 189

It is a lamp with an integrated switch. The lamp's structure is adjusted to interact with its user. It is assembled and disassembled in order to switch the light on and off, highlighting the users' intervention in the product and the products personality – making the light playful.

Materials: Aluminum

Page: Descriptions:

190 UNTITLED

Pick up the lamp and become part of it.

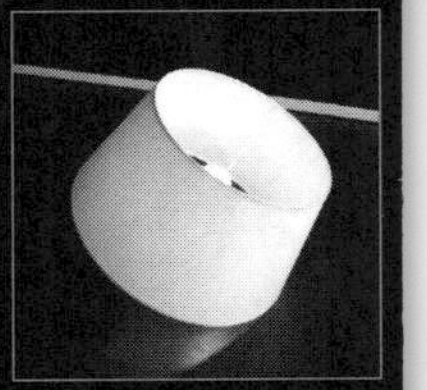

192 BALLOON LAMP

The designer wanted to make a global lamp that is simple and has minimum functions, with battery that doesn't need domestic electrical power. He hot hint from a Japanese lamp called 'CHOCHIN' 'CHOCHIN'. It is a portable lamp that is used with a candle in the EDO period. In his design, the flame is changed to led, the candle is changed to lithium cell and the paper shade is changed to rubber balloon. This lamp is made with balloon and led which has no generation off power like other lightings, such as incandescent lamps. And it has high illumination, long lifetime and is energy-saving. It is possible to use for over 100 hours continuously, with only 2 lithium-coin batteries.

194 CYCLOPS

Inspired by illogical placed and/or operating switches for lighting, the Cyclops design combines form with function, with it's minimalist yet elegant appearance incorporating innovative technology. Breaking the light by pulling 'the eye' downwards switches the lamp off, straighten it and it turns on; form is function. It is ideal for convenient office lighting above desks, but also suited for free hanging purposes in bars, living rooms, etc.

196 LIGHTABLE

Via 'total reflection', light from LEDs is transmitted through the glass top, which is not visible to human eye. But when placing an object that is capable to transmit light on the glass surface, neglecting the meeting surfaces with complete flush contact, light emits from the glass' surface and alights the object. It can be used as a shop front display of High-street retailers; department stores, perfumeries, etc. and create a more particular attention. Its potential could also be seen in its use for contemporary glass architecture: Each and every raindrop would have its trace so rain shower would create some remarkable fire works.

DEATHWISH MIRROR 198

Look mortality in the eye. A tribute to the miracle of death in any given context.

Deathwish is a custom mirrorized smoke black acrylic mirror.

SELFPORTRAIT (MIRROR SERIES) 199

The SELF-PORTRAIT mirror series puts a twist on the concept of vanity. Denouncing perceptions of negative self-image and self-criticism, these mirrors invite joy and humor by utilizing interactivity with the user's reflection.

Styles: Mustachio, Songbird, Nice Rack, Face Value, R, Pearl Necklace, The Draft, Face Value, Flock of Seagulls

PRETTY BIG WALL 200

Pretty Big Wallpaper is standard wallpaper with mirror elements as part of the pattern. Traditional ink patterns are combined with patterns made in mirror to create an almost 'live' composition that is unique to the space it reflects.

The mirrored elements are made by photo etching a thin sheet of mirror polish steel. The technology is simple and requires no investment in tooling which allows for customization and flexibility in production.

PRETTY BIG 201

Mirrors are often designed to free stand on a base or most commonly to be attached to a wall. This is mainly due to the use of traditional glass for the production of mirrors. Hanging Mirrors are made from mirror-finish, stainless steel; an unexplored material which is both lightweight and robust. This allows the objects to float in space either suspended from the ceiling or relying on the structure of other furniture. The iconic use of someone else's body as a surface to reflect our own image adds another dimension to the object, besides the functional.

Page: Descriptions:

202 MICOS

Micos is a collection of sculptural objects that challenge the rationale behind design for children and the culture it supports. Each abstracted object possesses a number of physical attributes which transform its character when its position is changed. It is designed with no specific function but, based on children's anthropometrics, they are reminiscent of furniture. They can be explored in a physical way, generating an understanding of the relationship between object and body, and the idea of use. This encourages multiple interpretations from the child.

204 HOTEL

The 'Hotel' box is neither a box, nor a piece of furniture, but all in between. It is a shoebox that is made of the finest quality paper ware from the UK. You buy them collapsed and sealed in plastic folio. All you do is to fold them into boxes and zip them with a pair of rubber bands. They can safely be stacked in up to six 10 towers fixed through a center screw at the top of the box and set up as shelves crawling along the walls by using some joints supporters, both vertically and horizontally. A dust lid is snapped into place through the opening in front, making this the dust-free window for display.

206 GRID WALL POCKETS

Ideal for storing magazines, remote controls, books, pens and other small items, Grid is a wall-mountable organizer with pockets. Made from die-cut and stitched recycled felt, this poster-size design is easy to install on any wall at home or in the office. With clean geometric patterns and vivid colours, Grid gives any setting a fresh contemporary look.

Philosophy: Grid Wall Pockets' geometric arrangements and use of vertical real estate for storage frees your desk, coffee table and even your toilet tank from unwanted clutter.

207 PRIME TIME

In the 18th Century the 'love seat configuration' where two people sat facing each other, was a very common piece of furniture. However, after the advent of television, the love seat slowly disappeared from mainstream furniture design.

Nowadays, almost all domestic seating is designed to face a television set, surround a dinning or a coffee table, thus physically and emotionally distancing ourselves. Prime time is designed as an one-off piece, that questions human relationships and the effect of technology in our lives.

Descriptions: Page:

ROCKING SLIPPERS 208

It is a hybrid clothing and furniture product creating a personal particular space. This project developed under the title of 'body extension' and the extension of the wearable till it gets almost to the territory of independent product. On the one hand, the designer wanted to explore the ability of human body to function differently with this extension, in this case the Rocking Slippers enable the body particular movement and posture and on the other she wanted to understand privacy within furniture.

GROWTH 210

A collection of domestic furniture inspired by patterns found in nature, making a poetic dialogue between the perception of the outdoor and the indoor. Inspired by the traditional way to decorate the houses interior with plants, this collection tries to apply the concept of 'bringing the nature to home' into furnitures that we can use daily. It is made by the repetition of an unic element. The repetition give them delicate tensions which are the key to make an structural and light furnitures. These pieces can also be used like outdoor furniture being a structure for the plants and giving another point of view for the landscapes.

SPLINTER 211

This teak and aluminum construction hints at a 'metaphor for a bad marriage day'. Consisting of two chairs that break apart or merge at a connection of jagged wood slats, this 'couple' also reunites as a complete bench. The visual approach to breaking up (with the irregular splintered ends) also allows one to sense the connection between the chairs from opposite ends of a room or patio. The chairs are designed for outdoor and indoor use.

MOLTO BEENE 212

A stool that bows – 10 legs provide flexible stability. Molto Beene plays with the necessary attributes of a seat: stability and comfort. Prototype made out of spring steel, linoleum and beech.

* Beene is Berlin pronunciation of legs.

Page:	Descriptions:

214 **BE YOUR OWN CHAIR ... AND WALK IT!**

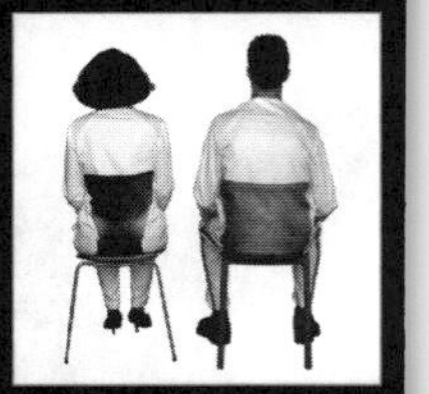

With our project Be your own chair ... and walk it! we have brought together the advantage of a simple traditional stool for milking cows with the elegance and the prestige of designer chairs. Aficionados of the icons of chair design can now furnish themselves with a classic - not just their home. Whoever slips into this suit becomes part of the myth! The chair as a second skin and its wearer are fused physically and psychologically breaking up the borders between human beings, furniture and space.

216 **BABUSCHKA**

Babuschka is a set of three differently sized chairs, which can be stacked into one another. Made from waxed birch plywood and painted on their insides, the chairs can be easily distinguished because of their different colours. In contrast to normal stacking chairs, it is possible to sit on the Babuschkas while they are stacked into each other.

218 **PREFAB DESIGN**

Prefabdesign is a furniture collection with designs that are sold as a DIY-kit. Due to the wide range of chairs (Decker), tables (Deck) and cupboards (Loader) the costumer can choose between many different designs. Because of the simple production method, the designs can be of low cost and accessible to everyone. Due to the wide range of furniture and the fact that it can be self-assembled, everyone can customize his own furniture. The website prefabfurniture.com shows three-dimensional manuals on how to assembly the products.

Little Decker: This baby-chair is part of Prefab design, the furniture collection with designs that are sold as a kit. All parts are kept together by wooden connections. After all connections are cut the parts can be painted. After this the little chair can be assembled. By the use of 2 screws and a bit of glue it can be fixed into one solid chair. Little Decker is accompanied by a little table: Little Deck.

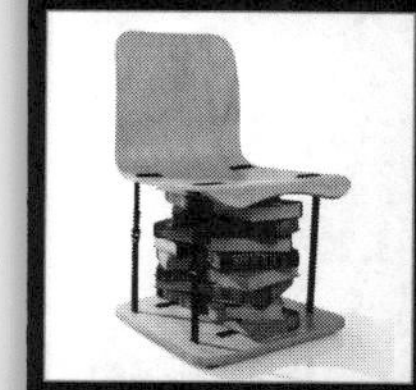

BALE CHAIR 220

Designed to re-use obsolete, seldom used or discarded stackable items such as encyclopedias, books, newspapers and magazines, the Bale Chair introduces do-it-yourself efficiency to the modern plywood aesthetic. Each Bale Chair consists of a molded plywood seat shell and plywood base made from responsibly harvested wood, nylon levelers and two recycled polyester webbing straps with steel buckles. To assemble this playful design, the selected items are stacked on the base to the desired height and are then secured with the webbing.

Philosophy: The Bale chair embodies the seamless integration of design and do-it-yourself ingenuity, nurturing an aesthetic that equally resonates with the simplicity of a studio or the boldness of a gallery. In adhering to the highest material environmental standards in the market today, the Bale Chair provides an unique experience of quality and performance. The design's playful approach to adaptive reuse invites users to actively participate in the design process.

TABLE CHEST 221

A chest of drawers can turned into a coffee table in a single action, by two sets of hinges. Made in Beech, Beech veneered MDF. Firstly self-production by the designer, then recently put back to European market by Roethlisberger.

LOVECHILD ROCKING STEP-STOOL 222

Function becomes questionable when an object is adorned with conflicting visual associations. Lovechild #1 marries a rocking chair and step ladder, whose true function is left to the user's discretion.

Page:	Descriptions:

223 MEMOIR

A table with an unfinished wood surface is produced to capture a moment in time. Celebrate! After a night of slurry conversations, unbreakable kinships and a few bottles (or boxes) of red wine. Memoir serves as a catalyst for recollecting fond memories from social gatherings of all kinds.

223 MEMOIR JR

A table with an unfinished wood surface is produced to capture a moment in time. Let the young'ns draw all over this piece of furniture over a period of days, weeks, months or years. When ready, the table is finished and sealed, yielding an object far more valuable to proud parents than any 'luxury' items in the house.

224 BENDABLE INTERIOR OBJECTS

The project 'Bendable Interior Objects' (B.I.O) is a futuristic vision. B.I.O is a concept built around the flat and slimmed production processes of today, where terms like packaging, assembly, transportation and environment are hot topics. B.I.O raises questions about our current living situation by delivering an unconventional and fresh furnishing concept.

B.I.O is laser cut in recyclable aluminium and comes in the shape of a flat "bag", easily carried home by the buyer. The user then snaps the objects out of the sheet metal and folds them into their intended shape by force of hand. By thinking B.I.O, we can save space, costs and environment.

Today the B.I.O concept consists of:
- The B.I.O Office kit which is a collection of three B.I.O products for the office (pen holder, letter holder and business card holder) contained in one A4 sized sheet
- The B.I.O coat hanger
- The B.I.O custom 'squeeze' hangers
- The B.I.O stool
- B.I.O Bin
- B.I.O Table clock

The B.I.O products are sold in design shops over the globe; Japan, Hong Kong, Singapore, Greece, Germany, Switzerland, France, Sweden, Italy, UK, USA, Mexico, etc.

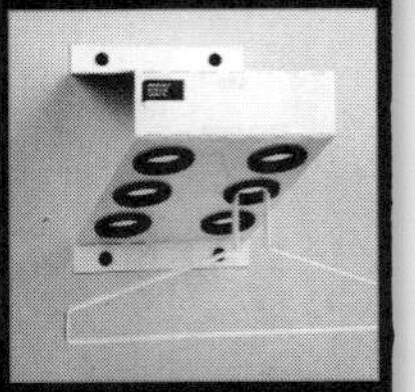

GROUP OF TREES 228

The future is flexible! Group of Trees is a flexible and adjustable sound-absorbing room divider for public spaces.

Room dividers and sound absorbers are typically flat, geometrical and clumsy. Group of Trees is organic, flexible and interactive. This is a completely new way of looking at room dividers and sound absorbers. Group of Trees is adjustable in both width and height. The modules, made out of a sound absorbing, pressed polyester felt, are easily attached to one another without any tools through a pin-fitting device. A cylinder moulded into the base allows a 360-degree rotation so that the user can choose the level of visual and aural screening. The entire room divider is mobile and encourages interaction between user and product. Group of Trees is a result of the study of creative environments and how our surroundings affect how we learn and create. A flexible and inspiring environment encourages new ideas and new ways of thinking.

The product fits public environments such as libraries, museums and office spaces, where large rooms and large numbers of people create much noise. The modules can be fixed in their positions or they can be left open for the visitor, user or employee to modify in height and width if they, for example, need to create a small room inside a bigger one for a private meeting. This kind of product is needed on the market. There is a product gap between traditional sound absorber walls and modern-looking room dividers. The product consists of two parts – a base and 'wings', which makes it easy to produce, package, store and transport. The process of pressing heated polyester felt with glue fibres is established and common in the making of interiors for vehicles.

The felt 'wing' is made out of two similar sides, which requires only one press tool. The two different trees are made of the same type of wings that are placed in different positions. The two sides of the wing are joined together with glue and the core is left empty for increased sound absorbing qualities. The steel plate base is lasercut. Group of Trees is flexible and easy to modify. The wings are lightweight and easy to dismantle, and can also be pressed with handles to aid interaction.

SWOPIT 229

Swopit is a chair and table combination of three-dimensional furniture with two-dimensional graphics. The concept is built around the idea that the users are able to change the whole character of the furniture in a couple of seconds. A textile, pattern, paper, or whatever you prefer goes in between the two transparent acrylic layers and gives the chair and table the desired look. Only the users own imagination sets the limits.

Page: Descriptions:

230 COFFEECUP

With the idea of 'Drink it like Beckham' (The next big soccer event is coming!), CoffeeCup turns the IKEA table 'Lack' into a field of glory, guaranteeing every half sweet shots and crossbar collutions.

Instruction: 1. Saw out the cutting and drill the holes at the sites of the IKEA table 'Lack'; 2. Shorten the table soccer poles in the centre and push them through the table; 3. Fix the sugar dispenser and the can of milk.

231 TURNTABLE

'Only the stupid need organisation, the genius controls the chaos.' Albert Einstein. But at times even the genius longs for an empty tabletop. The Turntable helps us control the chaos. Without distroying the beloved and creativity aiding disorder, the Turntable stores it away within itself*, easily to be restored at any time.

* Hereby the disorder is actually preserved in it's original place, simply the space is being warped.

Material: Felt, Steel, Wood / Colour: White, Charcoal.

232 YUKARI-MASU

'yukari-masu = 縁枡' has the mind of 'relation = 縁', and the 'square = 枡'. Each 'square' has a meaning when it is used; and when they are used together, it is like building 'relationship' between people.

233 MAGNETISCH / MAGNETIC TABLE

Magnetic table is a table system. Individual work, conference, kitchen or dining tables can be generated. The whole bottom of the tabletop is covered with a magnetic foil. All parts of the kit adhere underneath the tabletop. The tabletop is non-magnetic. Effects on electronic equipment are impossible.

Descriptions: Page:

EXPECT CABINET 234

Cabinet designed after a research on experiences and expectations on interior products.

REBELLIOUS CABINET I 235

The drawers of this cabinet can never be closed all at the same time. When closing one drawer, the one next to it will open a bit. If you want to open a drawer that is fully closed, you just hit, kick or push the drawer next to it.

'GREEN DAY' SERIES 236

The inspiration is mainly come from the designer's feeling of 'I want to walk on grass with my bare foot, in rainy days, in winter... I want to release from our fast track of daily life.' The lawn 'green day' is made with a material called Pseudofabric from grass, a kind of paper yarn OJO+ that can be broken down into carbon dioxide and water in the soil when discarded. The material used is the strictest condition of design. The designer has been keeping himself up-to-date of environmental material usage instead of just having a good-looking product design.

SIZE? (I.E.: NOAILLES, BELT & SAMPLE) CARPET 238

This series of 3 carpets is the result of collaboration with a French company Tisca. Adrien Rovero continues his exploration of rugs after his carpet 'Dis-Order' (Diplomas work at Ecal in 2004). It interrogates the size and the shape of usual rugs (square and available in one dimension). They are all based on a connection detail never applied to flooring: a chain, a belt and a sample piece. These assembly principles generate new forms as well as custom sizes and keep the traditional hand tufted techniques. It is made of 100% Wool.

Page:	Descriptions:

240 ZIPZIP

ZipZip are upholstered cushions that can be connected with zips on all four sides. Countless possibilities to build nearly every situation of sitting or laying down comfortably are the result. The width of a sofa just depends on how many polsters are connected horizontallly. If they are folded 'zigzag'-like, the hight of the seats can be varied. If turned to the wall, a backrest is ready to use. Armchairs, couches, a carpet or a bed come into being within the twinkle of an eye. Special situations of rooms can be included in awareness with ZipZip. It is easy and flexible to use.

242 FILLING THE CAT

Three characters disappear from large, solid white blocks, and their absence suddenly makes them important. The resulting emptiness is filled with the user's belongings and through this action the silhouettes' meaning shifts. The pieces suggest a continuous play between interaction and representation, where daily use generates a continuously changing story: a cat that reads Italo Calvino, a book filled with coins, a living room where birds come to eat the cake's crumbles.

246 IVY

Ivy is like lingerie for the wall. An adventurous decoration that works as a free-climbing wall to see your 'over designed' interior from a different perspective or find alternative routes to go upstairs. Ivy is also an ode to freedom and meaning. It is made out of 7 different pieces that can be combined in every single way. Ivy is made out of polyester concrete which is a mixture between silver sand and polyester. A material normally used for sports climbing grips. And like these the Ivy pieces are attached to the wall with Allen-screws.

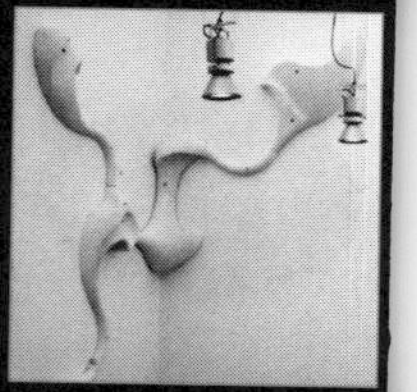

248 CERAMIC CACTUSES

It is an intuition piece part of Jaime Hayon's visual world.

Title
AFTER DARK
Dimension
VARIOUS
Design / Manufacture
AFRODITI KRASSA LTD
Photo
FRANK THRUSTON
Client
PA
Quantity
100
Awards
TROIKA DESIGN AWARDS
Year
2002

> See also:
Page 068-069, 200-201, 207

Title
SOLITAIRE DISH 1
O'S AND X'S ASHTRAY 2
BATTLESHIPS NAPKINS 3

Dimension
21 X 21 X 4 CM 1
10 X 10 X 3.5 CM 2
12.5 X 12.5 CM 3

Design
ANDRÉ KLAUSER

Client / Manufacture
THORSTEN
VAN ELTEN, UK

Quantity
3000+ 1
500+ 2
100,000+ 3

Collection
DESIGN MUSEUM
LONDON SHOP,
MOMA NEW YORK SHOP 1

Year
2003

> See also:
Page 171

solitaire ©thrink

Before play commences, 32 olives are placed in the small dips, except for the one in the centre.

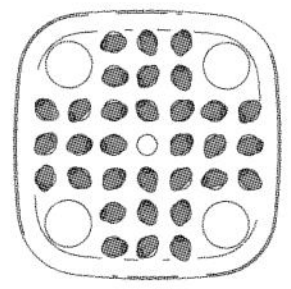

The objective of the game is to remove(and eat) the olives until only one remains.

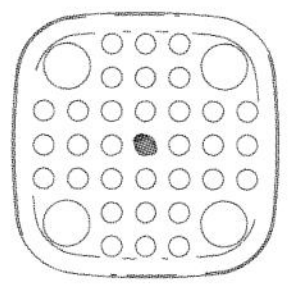

An olive can be removed only when an adjacent olive jumps over it. It can only jump into an empty dip and only over one olive at a time. Diagonal moves are not allowed.

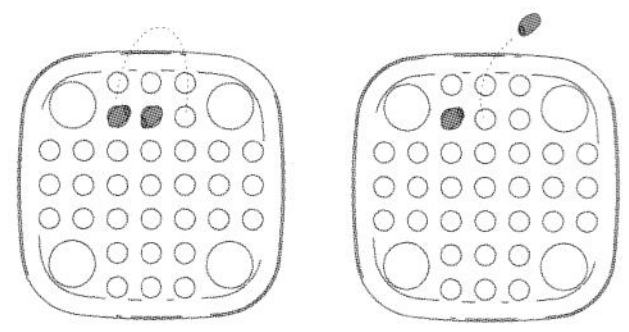

The pips are discarded in the four large dips provided. Olives can be replaced with nuts or any similar snack. Bon Appétit.

o's and x's ©thrink

Also known as tic-tac-toe, o's and x's is a two player game. The players take it in turn to leave their cigarette end in one of the nine dips in the ash tray.

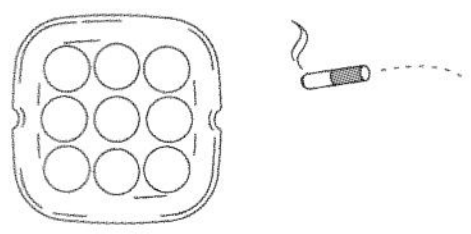

The game is won when a player manages to place three of his cigarette ends in a line. This can be from left-to-right, top-to-bottom or diagonally.

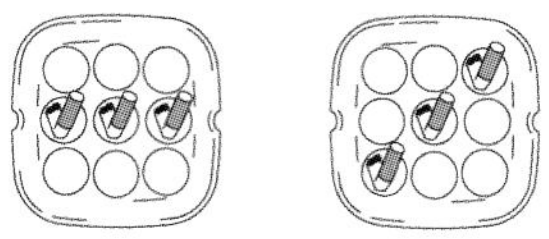

If neither player can achieve this and there is no dip left in the ash tray then the game is drawn.

Alternatively to cigarette ends, the dips can be marked with flicks of ash.

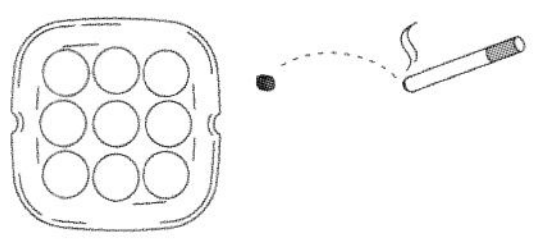

battleships napkin ©thrink

A game for 2 (1 napkin per player)

1.Mark the location of your ships on the napkin. Only use areas free of land.

2.Take turns to shoot at your opponents ships by calling the coordinates you are aiming for.

3.Hits and misses must be called. Mark your hits and misses and your opponents hits by using x and o.

4.A player who lands a hit can strike again. Sink the opponents fleet before he can sink yours.

www.thorstenvanelten.com

Title
"DO DESTROY" GIFT CARDS
Design
NADINE JARVIS
Year
2003

> See also:
Page 172, 175

Title
STICKYSTAINS
Dimension
11.5 X 16 CM
Design / Client
SHAY ALKALAY
Quantity
1000
Year
2006

> See also:
Page 126-127, 176-177

Title

GAS BAGS

Design

NADINE JARVIS

Year

2004

> See also:
Page 172-173

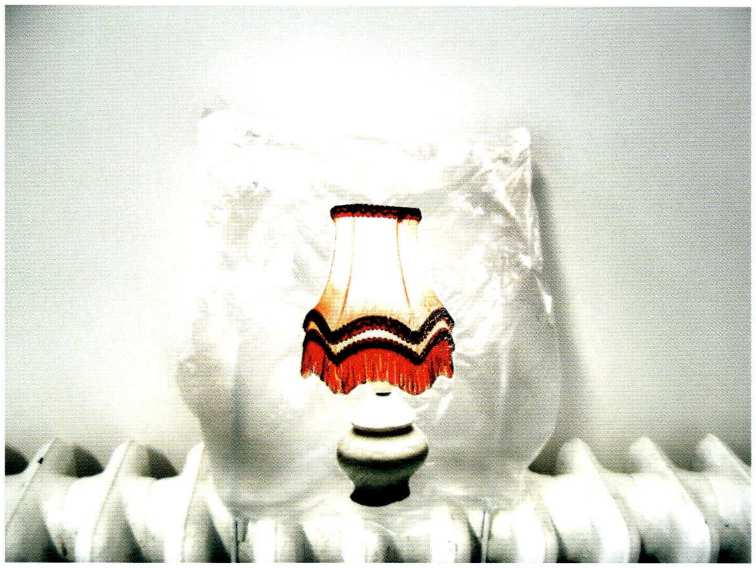

Title
HOOVER BAG BEARS 1
BIN BEAR BAG 2

Dimension
20 X 12 X 12 CM 1
50 X 50 X 50 CM 2

Design / Client
SHAY ALKALAY

Quantity
10 1
100 2

Collection / Awards
TALETED 2006,
BRITISH COUNCIL 2

Year
2006

> See also:
Page 126-127, 174, 177

1

2

Title
PRIVACY SCARF
Dimension
VARIOUS
Design / Photo / Manufacture
JOE MALIA
Client
RCA
Quantity
PROTOTYPE
Year
2006

> See also:
Page 179

Title
PUNCH'N'CUDDLE BAG
Dimension
160 X 70 CM
(COMMISSIONS UNDERTAKEN)
Design / Manufacture
SHOOT THE STYLIST!
Client
COMMISSIONS UNDERTAKEN
Quantity
PRODUCE ON REQUEST
Awards
PRINCESS OF WALES AWARD
Year
2006

> See also:
Page 180

Title
PLANT TAPE
Dimension
5 X 600 CM
Design
CHRIS KABEL DESIGNS
Quantity
LIMITED
Year
2002

> See also:
Page 058-059, 062-063,
084-085, 183

Title
GREEN LETTUCE
Dimension
22 X 16 X 13 CM
Design / Manufacture
KARIN VAN LIESHOUT
Photo
RUY TEIXEIRA
Client
VARIOUS
Quantity
25
Year
2001

> See also:
Page 054-055, 090-091, 185

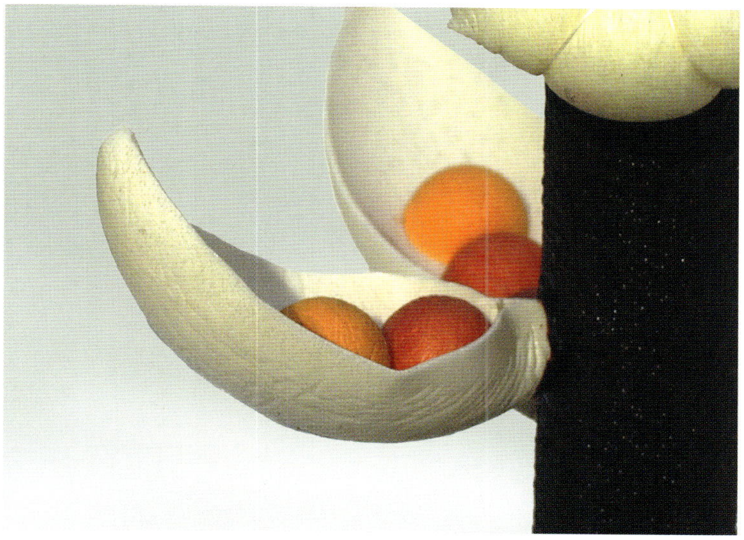

Title
CAPILLARY TREE
Dimension
16 X 16 X 45 CM
Design / Photo / Manufacture
KARIN VAN LIESHOUT
Quantity
PROTOTYPE
Year
2001

> See also:
Page 020, 130, 212-214, 254

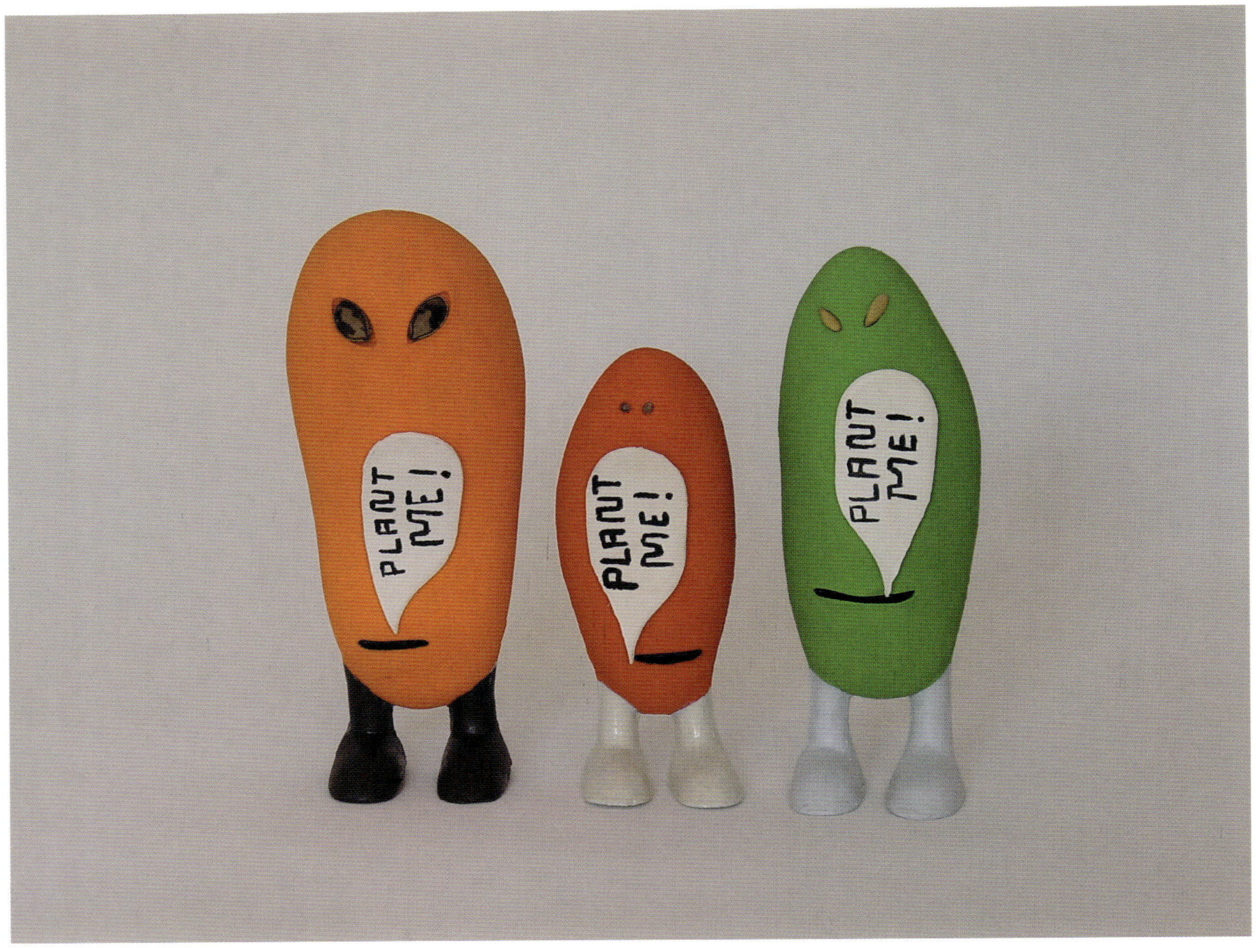

Title
PLANT-ME PET
Dimension
6 X 20 CM
Design
CONCEPTS AND IDEAS FOR
COMMERCIAL PURPOSES S.L.
Photo
IMAGEKONTAINER / INGA KNÖLKE
Client / Manufacture
THE ORIGINAL CHA CHA, BARCELONA
Quantity
MASS PRODUCTION
Awards
DESIGN PLUS 2004 AWARD
Year
2003

> See also:
Page 136-137

Title
HONEYCOMB LAMP
Dimension
45 X 30 CM
Design / Photo
KYOUEI DESIGN
Client
CONRAN SHOP NY,
ROPPONGI-HILLS TOKYO, ETC.
Manufacture
KYOUEI DESIGN, NAGAMASA
PAPER MANUFACTURE
Year
2005

> See also:
Page 192-193

Title
JELLYFRIENDS
Dimension
40 X 25 CM DIAMETER
Design
FORM US WITH LOVE®
Photo
FREDRIK GRANVIK
Client / Manufacture
UNO DESIGN (SPAIN)
Year
2006

> See also:
Page 224-229

Title
ON OFF
Dimension
17 X 65 CM
Design / Photo / Client / Manufacture
REDDISH
Quantity
PROTOTYPE
Awards
THE MINISTRY OF INDUSTRY, TRADE AND LABOR AWARD FOR INDUSTRIAL DESIGN (CONCEPT DESIGN CATEGORY), ISRAEL
Year
2002

> See also:
Page 072-073, 082-083

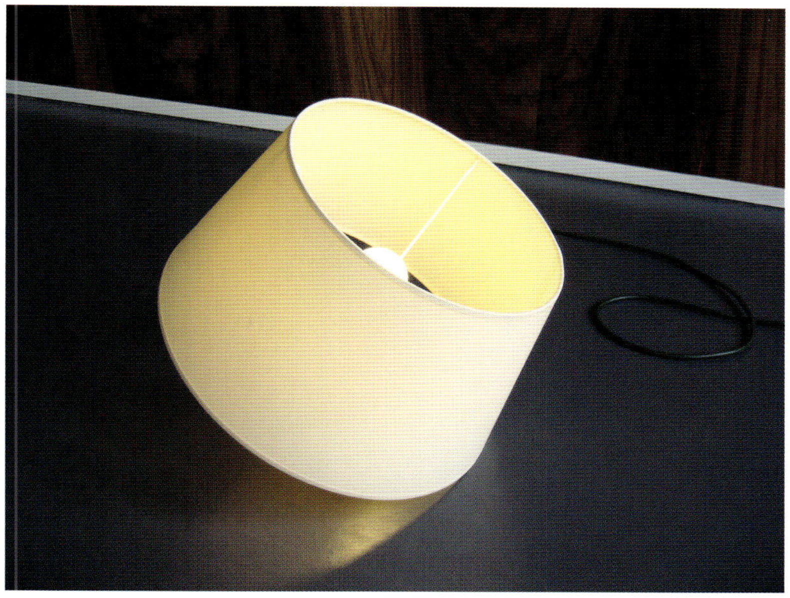

Title
UNTITLED
Dimension
VARIOUS
Design
HELMUTSMITS.NL
Year
2006

> See also:
Page 190

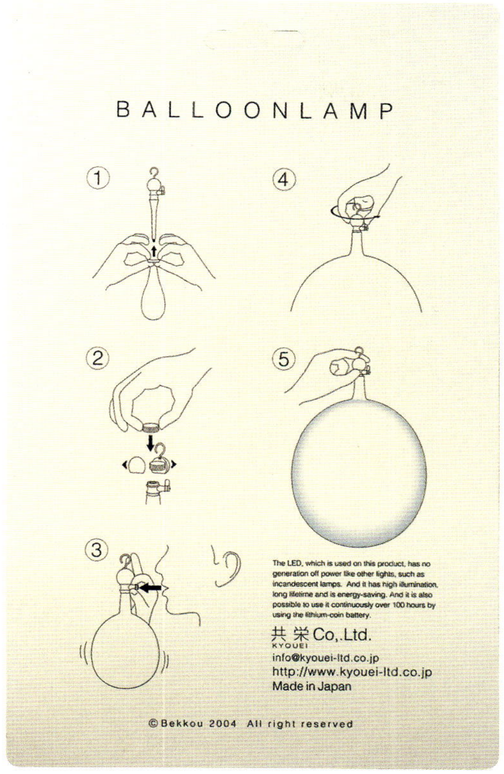

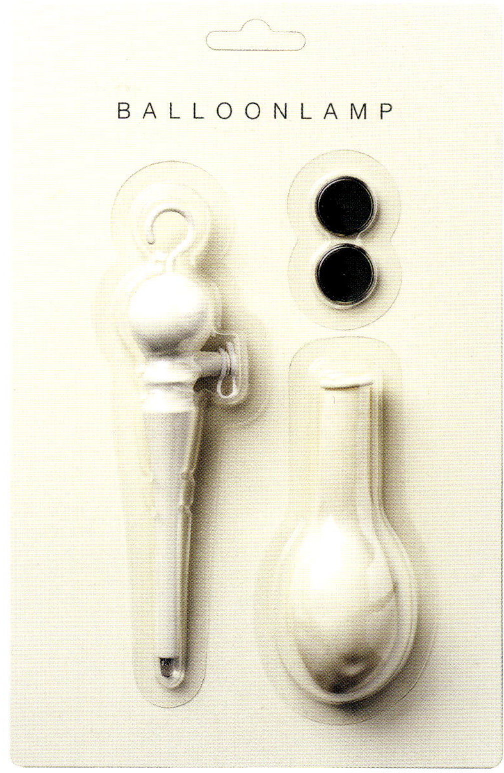

Title
BALLOON LAMP
Dimension
15.5 X 4 CM (MAIN BODY)
Design / Photo
KYOUEI DESIGN
Client
DROOG, PAUL SMITH,
ROPPONGI-HILLS TOKYO, ETC.
Manufacture
KYOUEI DESIGN, TAMIYA LTD.
Year
2003

> See also:
Page 187, 192

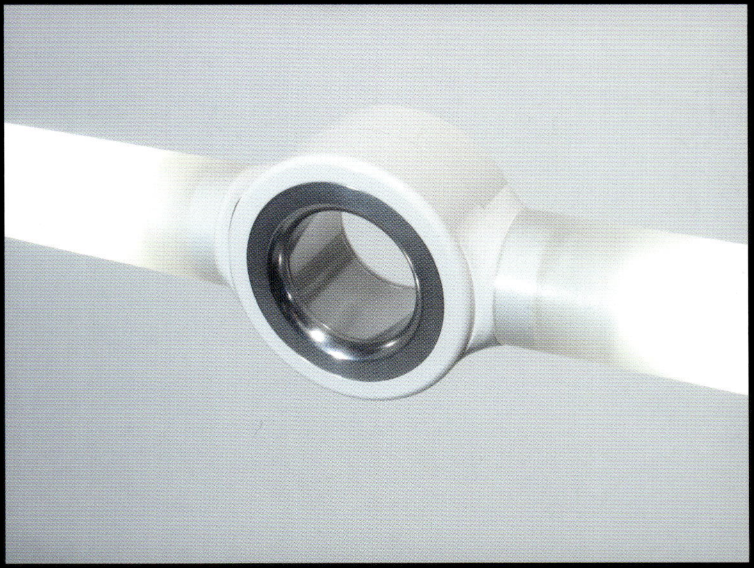

Title
CYCLOPS
Dimension
130 X 8 X 4 CM
Design
BURO VORMKRIJGERS
Photo
SANDER MULDER

> See also:
Page 044, 120, 122-123,
135, 194, 218-219

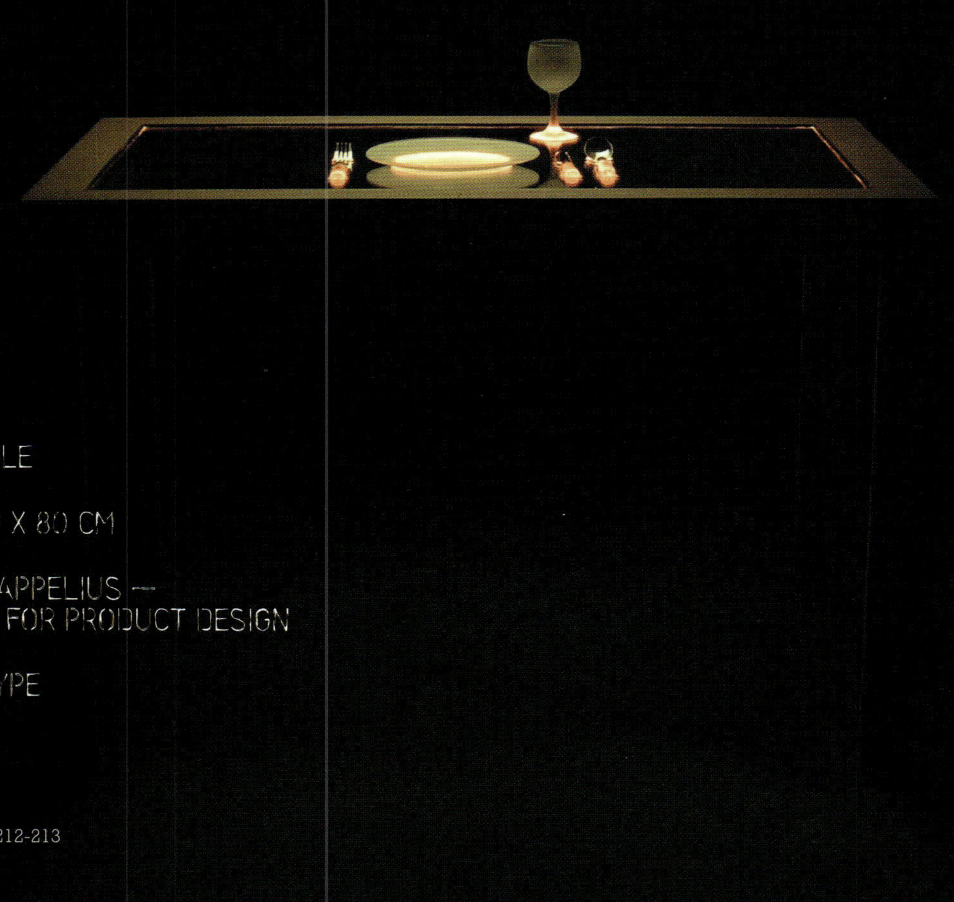

Title
LIGHTABLE
Dimension
140 X 70 X 80 CM
Design
JULIAN APPELIUS —
BUREAU FOR PRODUCT DESIGN
Client
PROTOTYPE
Year
2001

> See also:
Page 196, 212-213

Title
DEATHWISH MIRROR
Dimension
36.8 X 40.6 X 0.6 CM
Design / Client / Manufacture
THE DESIGN CAN
Quantity
120+ (IN PRODUCTION)
Year
2006

> See also:
Page 199, 222-223

Title
SELF PORTRAIT
(MIRROR SERIES)
Dimension
30.5 X 40.6 CM
Design / Client / Manufacture
THE DESIGN CAN
Quantity
1800+ (IN PRODUCTION)
Year
2004

> See also:
Page 198, 222-223

1

2

Title
PRETTY BIG WALL 1
PRETTY BIG 2

Dimension
200 X 300 X 1.2 CM 1
220 X 46 X 0.1 CM 2

Design / Manufacture
AFRODITI KRASSA LTD

Photo
EMMA THOMPSON

Client
THE DIVINE COLLECTION

Quantity
12 1
60 2

Year
2004 1
2003 2

> See also:
Page 068-069, 169, 200, 207

Title
MICOS
Dimension
60 X 65 X 35 CM
Design / Manufacture
EL ULTIMO GRITO
Photo
GARETH GARDNER
Quantity
LIMITED EDITIONS
Year
2006

> See also:
Page 203

Title
HOTEL
Dimension
15 X 47 X 30 CM
Design / Client
PURE NOMADE
Quantity
14000
Year
2006

> See also:
Page 205

Title
GRID WALL POCKETS
Dimension
68 X 53 X 1.3 CM
Design / Client / Manufacture
MIO
Photo
ROBERT HAKALSKI
Year
2003

> See also:
Page 220

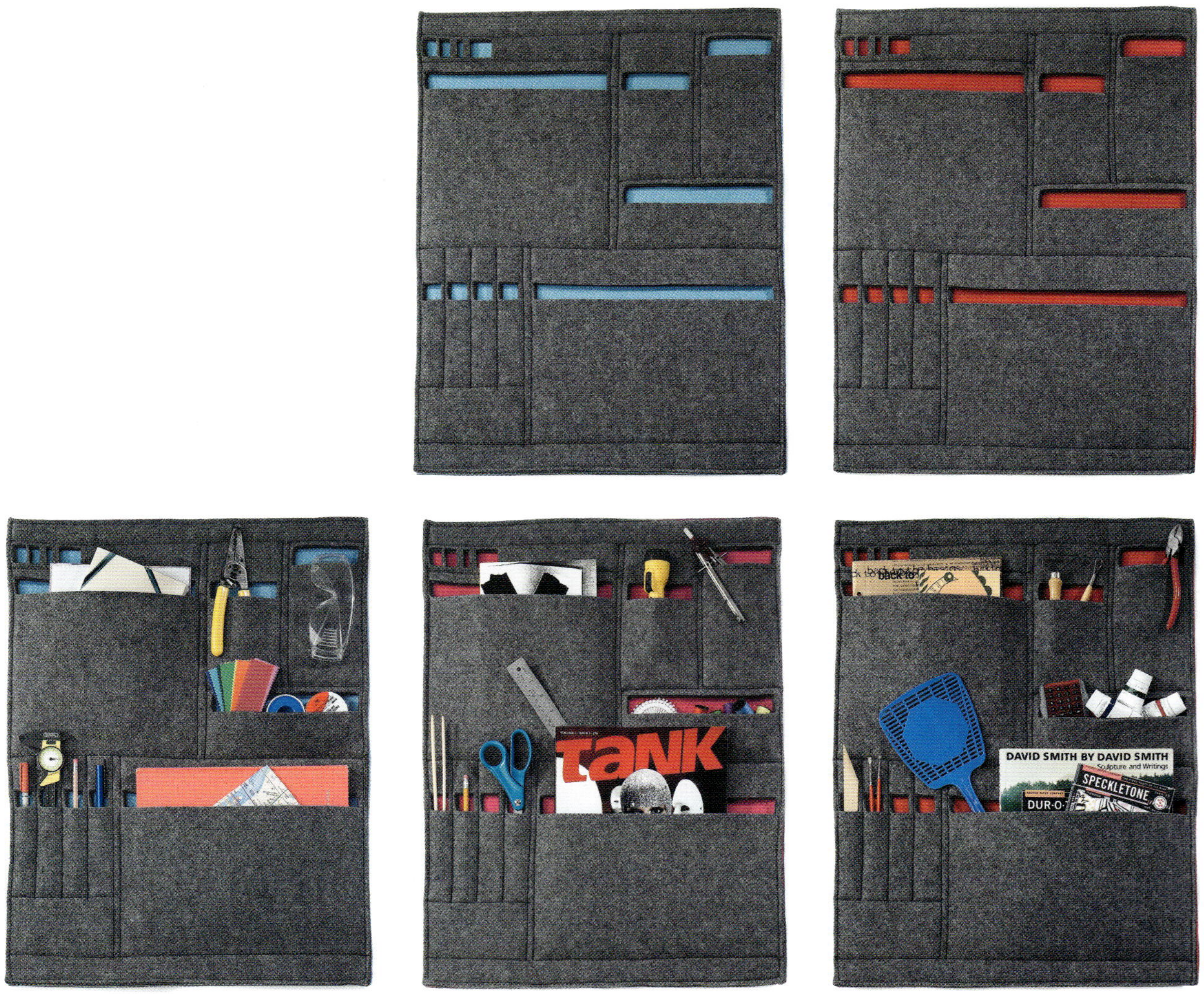

Title
PRIME TIME
Dimension
120 X 80 X 76 CM
Design / Manufacture
AFRODITI KRASSA LTD
Photo
FRANK THURSTON
Client
HIDDEN
Quantity
PROTOTYPE
Year
2001

> See also:
Page 068-069, 169, 200-201

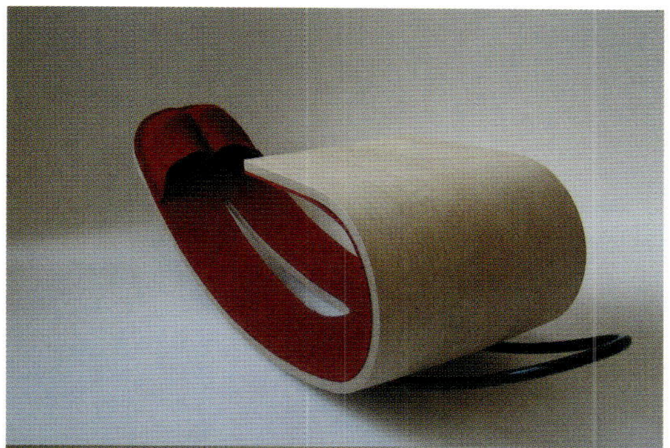

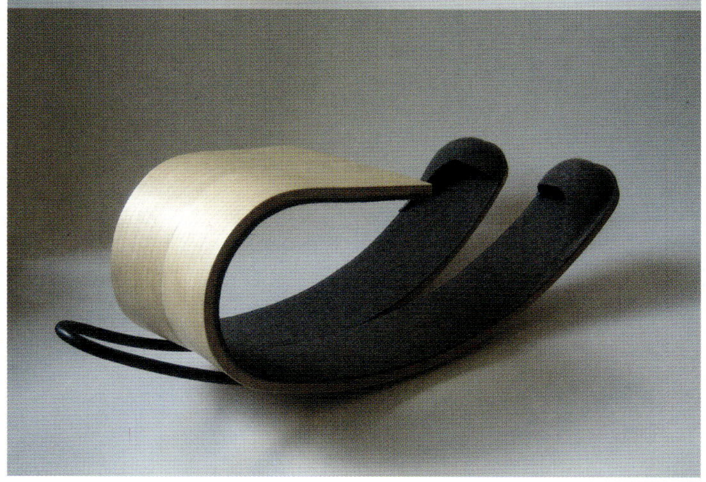

Title
ROCKING SLIPPERS
Dimension
75 X 30 X 22 CM
Design
YAEL MER
Photo
BRIT LEISSLER
Quantity
PROTOTYPE
Year
2006

> See also:
Page 110-111, 208

Title
GROWTH
Dimension
120 X 50 X 60 CM,
60 X 50 X 65 CM,
125 X 80 X 75 CM
Design / Client / Manufacture
ELI-GUTIERREZ STUDIO
Photo
FRAN AVILA
Quantity
3
Awards
SELECTED FOR
THE IM COLOGNE 2007
Year
2006

Title
SPLINTER
Dimension
76.2 X 137.2 X 55.9 CM
Design / Manufacture
MATTHEW KROEKER
INDUSTRIAL DESIGN
Photo
TIM POLNAU
Quantity
PROTOTYPE
(FULL PRODUCTION IN 2007
WITH JANE HAMLEY WELLS)
Awards
BEST PROTOTYPE 2006,
METROPOLIS MAGAZINE AWARD
(AT THE INTERIOR DESIGN
SHOW IN TORONTO CANADA)
Year
2006

Molto Beene

Title

MOLTO BEENE

Dimension

50 X 28 X 48 CM

Design

JULIAN APPELIUS — BUREAU FOR PRODUCT DESIGN

Quantity

PROTOTYPE

Year

2006

> See also:
Page 196-197, 212

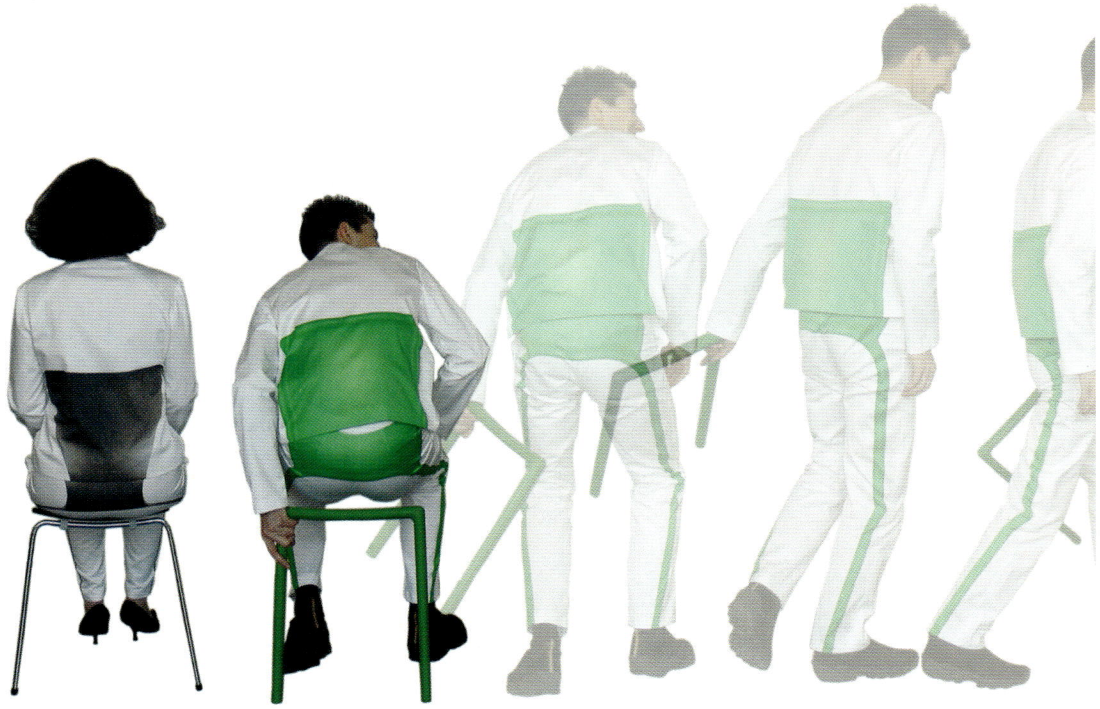

Title

BE YOUR OWN CHAIR ...
AND WALK IT!

Dimension

FOUR "SEATS", EACH APPROX.
48 X 45 X 8 CM, FOUR SUITS

Design / Photo / Client / Manufacture

STAUFFACHERBENZ

Quantity

4 DIFFERENT SETS
(CONSISTING OF SHIRT, TROUSER
AND "SEAT"), ALL ARE UNIQUE PIECES

Awards

DESIGN PREIS SCHWEIZ 2003

Year

2003

> See also:
Page 128, 214

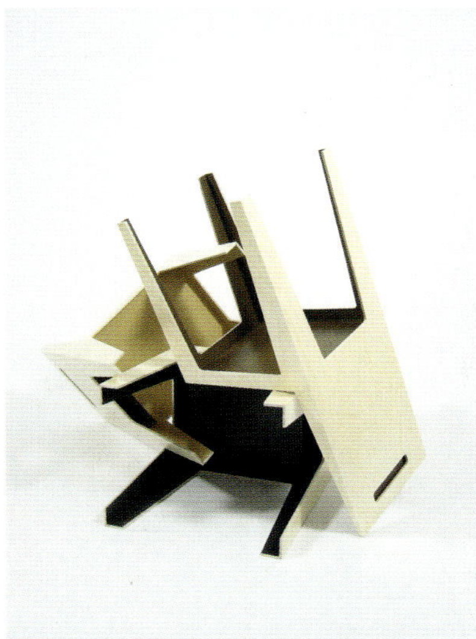

Title
BABUSCHKA
Dimension
BIGGEST CHAIR :
36 X 46 X 87.5 X 46.5 CM
(OTHER CHAIRS 1 CM
SMALLER IN EVERY
DIRECTION)
Design
STERNFORM
PRODUKTGESTALTUNG,
B.KNUEPPEL
Photo
STERNFORM
Manufacture
CONCAVEX GMBH

Awards
1ST PRIZE, HABITAT DESIGN
COMPETITION 2003
Year
2003

> See also:
Page 010-011, 028-029,
034-037, 216, 240-241

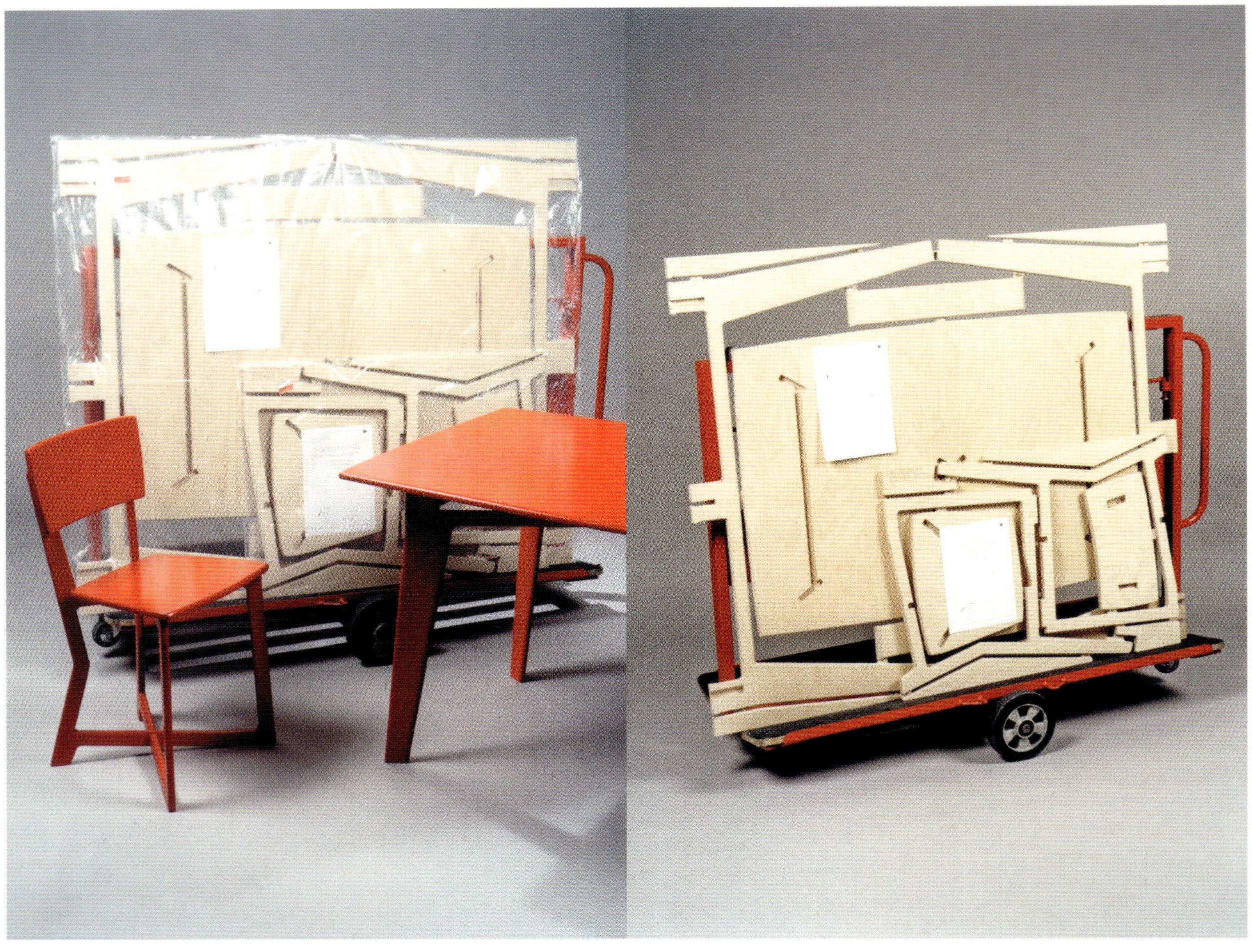

Title
PREFAB DESIGN
Dimension
DECKER : 81 X 40 X 43 CM
DECK : 74 X 80 X 120 CM
LITTLE DECKER : 55 X 30 X 32 CM
Design
BURO VORMKRIJGERS
Photo
DAVE KEUNE
Awards
NOMINATED FOR ITEMS AWARD

> See also:
Page 044, 120, 122-123,
135, 194-195, 219

Title
BALE CHAIR
Dimension
68 X 68 X 68 CM
BASE : 48 X 46 CM
Design / Photo / Client / Manufacture
MIO
Year
2004

> See also:
Page 206

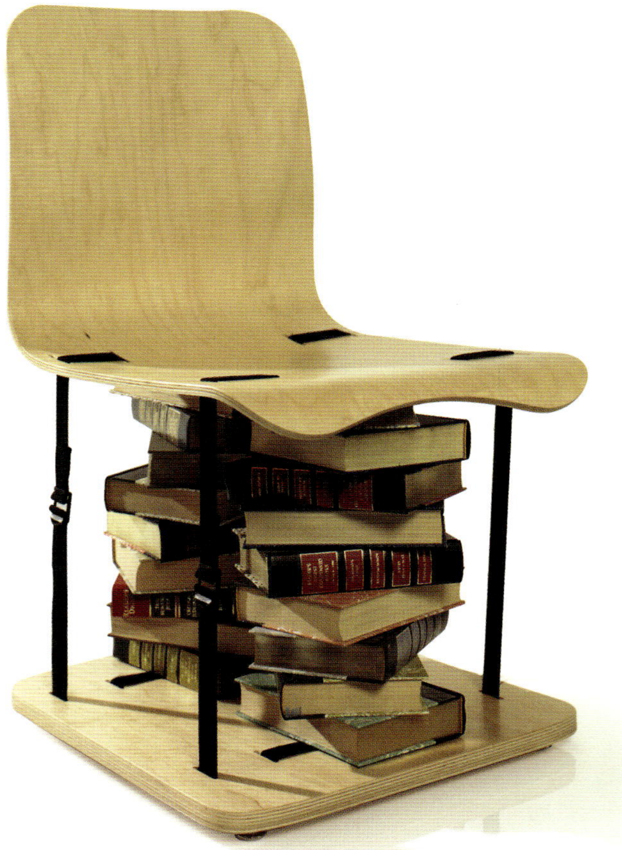

Title
TABLE CHEST
Dimension
40 X 40 X 75 CM
Design
T.N.A. DESIGN STUDIO
Photo
YASUHIRO SAKURAI
Manufacture
ABODE (JAPAN), ROTHELISBERGER (SWITZERLAND)
Quantity
APPROX. 50
Collection
VICTORIA & ALBERT MUSEUM (LONDON), GEFFRYE MUSEUM (LONDON), BRIGHTON MUSEUM (BRIGHTON), MANCHERSTER ART GALLERIES (MANCHESTER), CRAFTS COUNCIL (LONDON)
Awards
FX NEW DESIGNERS AWARD 1995
Year
1995

tdc

Title

LOVECHILD ROCKING STEP-STOOL 1
MEMOIR 2
MEMOIR JR 3

Dimension

45.7 X 61 X 66 CM 1
76.2 X 152.4 X 76.2 CM 2
71.1 X 71.1 X 50.8 CM 3

Design / Client / Manufacture

THE DESIGN CAN

Quantity

4 (FURTHER DEVELOPMENT AND MANUFACTURING IN PROGRESS) 1
1 (PRODUCE ON REQUEST) 2
4(PRODUCE ON REQUEST) 3

Year

2006 1,3
2004 2

> See also:
Page 198-199, 222

2

3

BENDABLE™
INTERIOR
OBJECTS

Title
BENDABLE INTERIOR OBJECTS (COAT HANGER, SHEET, STOOL, OFFICE, ETC.)

Dimension
0.2 CM SHEET ALUMINUM

Design
FORM US WITH LOVE®

Photo
JONAS LINDSTRÖM

Client / Manufacture
BENDABLE INTERIOR OBJECTS

Quantity
1000+

Awards
YOUNG SWEDISH DESIGN 2006

Year
2005/6

> See also:
Page 188, 224, 226-229

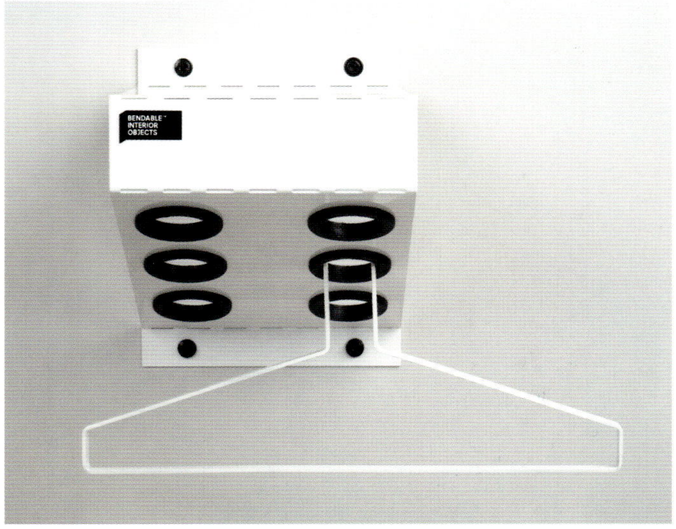
BENDABLE
INTERIOR
OBJECTS

BENDABLE
INTERIOR
OBJECTS

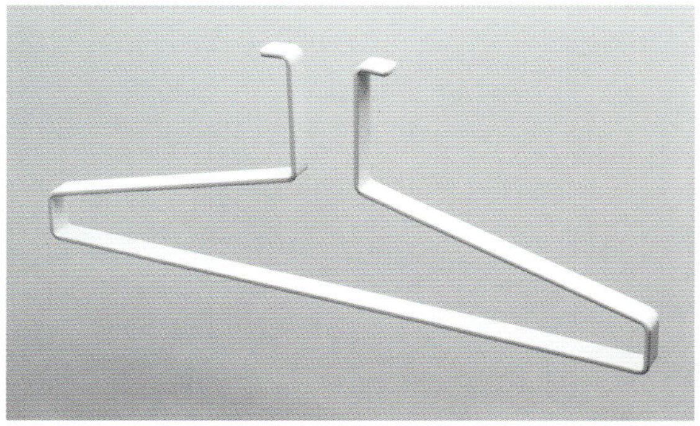

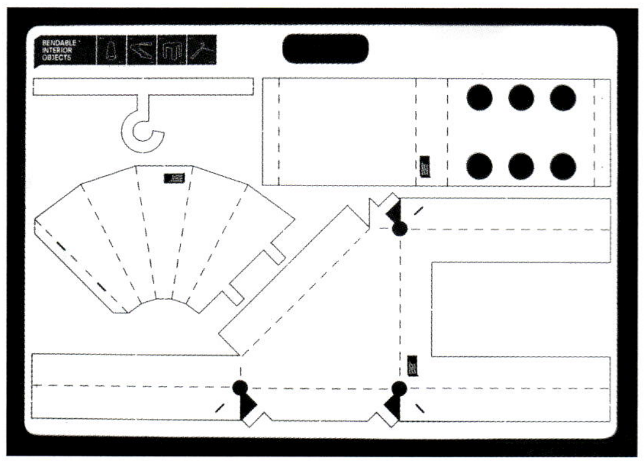

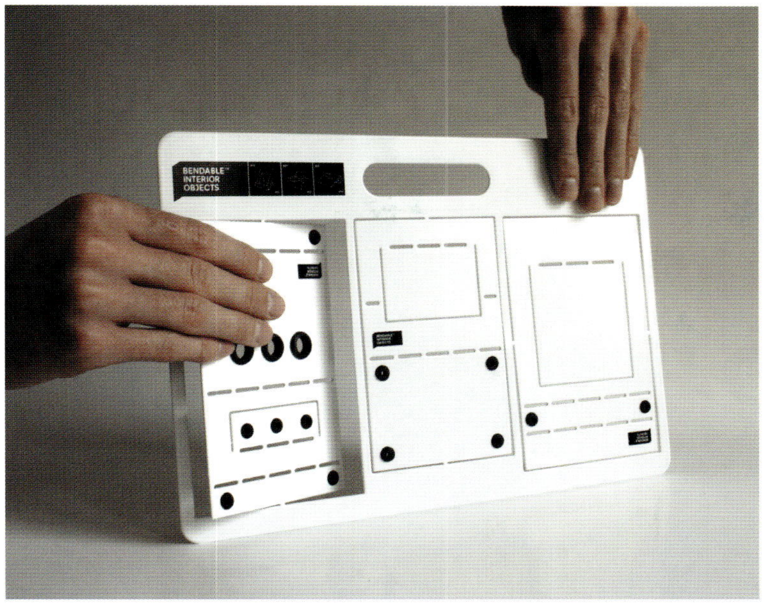
BENDABLE
INTERIOR
OBJECTS

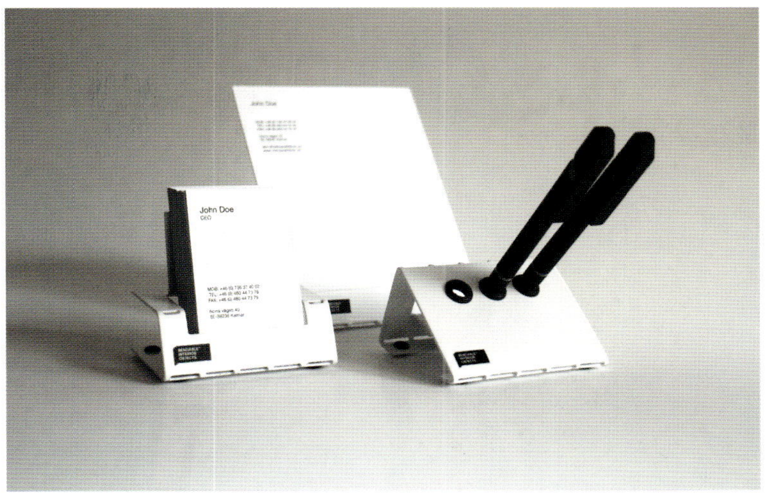
John Doe

Title
GROUP OF TREES
Dimension
170 X 40 X 10 CM
Design
FORM US WITH LOVE®
Photo
JONAS LINDSTRÖM
Client / Manufacture
MATERIA
Awards
RED DOT DESIGN AWARD 2006
Year
2006

> See also:
Page 188, 224-227, 229

Title
SWOPIT
Dimension
40 X 30 X 50 CM
Design
FORM US WITH LOVE®
Quantity
PROTOTYPE
Year
2005

> See also:
Page 188, 224-228

Title
COFFEECUP
Dimension
90 X 55 X 45 CM
Design / Photo / Manufacture
:: DAN PEARLMAN
Quantity
2
Collection
FUSSIDELUXE
Year
2004

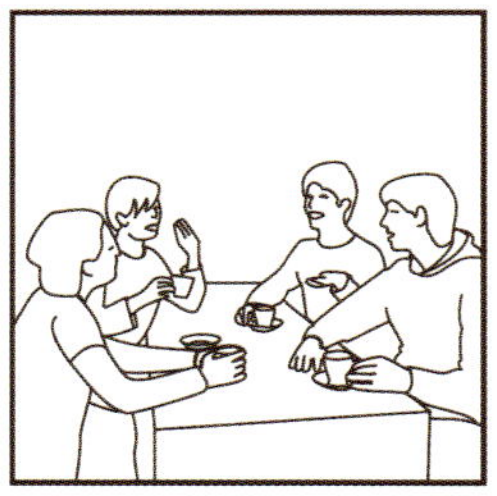

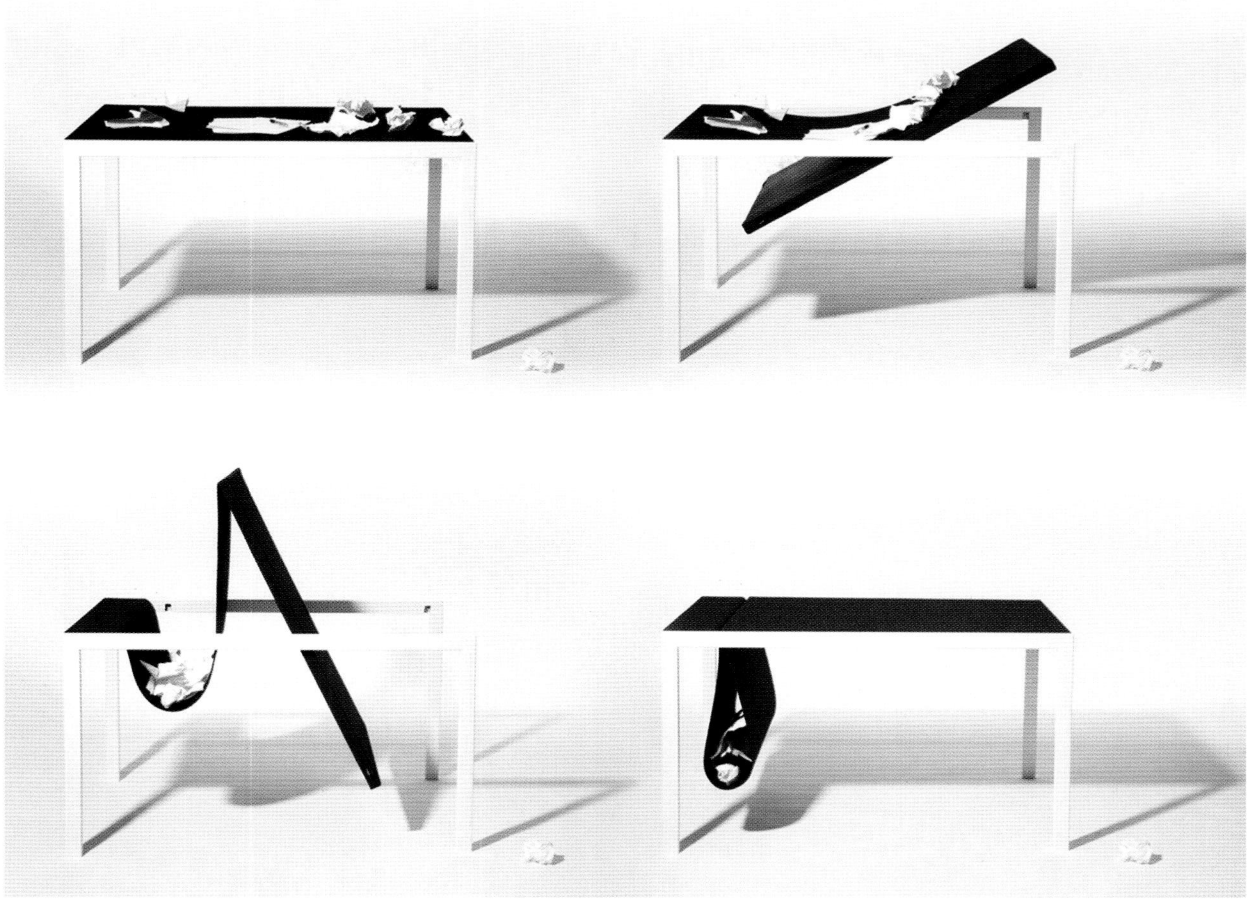

Title
TURNTABLE
Dimension
130 X 80 X 75 CM
Design
FEINSTET
Awards
ITALIAN GIOA CASA —
100 BEST DESIGN PRODUCTS 2006
Year
2006

Title

YUKARI-MASU

Dimension

44.5 X 44.5 X 8 CM

Design / Client

PD DESIGN STUDIO

Year

2000

> See also:
Page 236-237

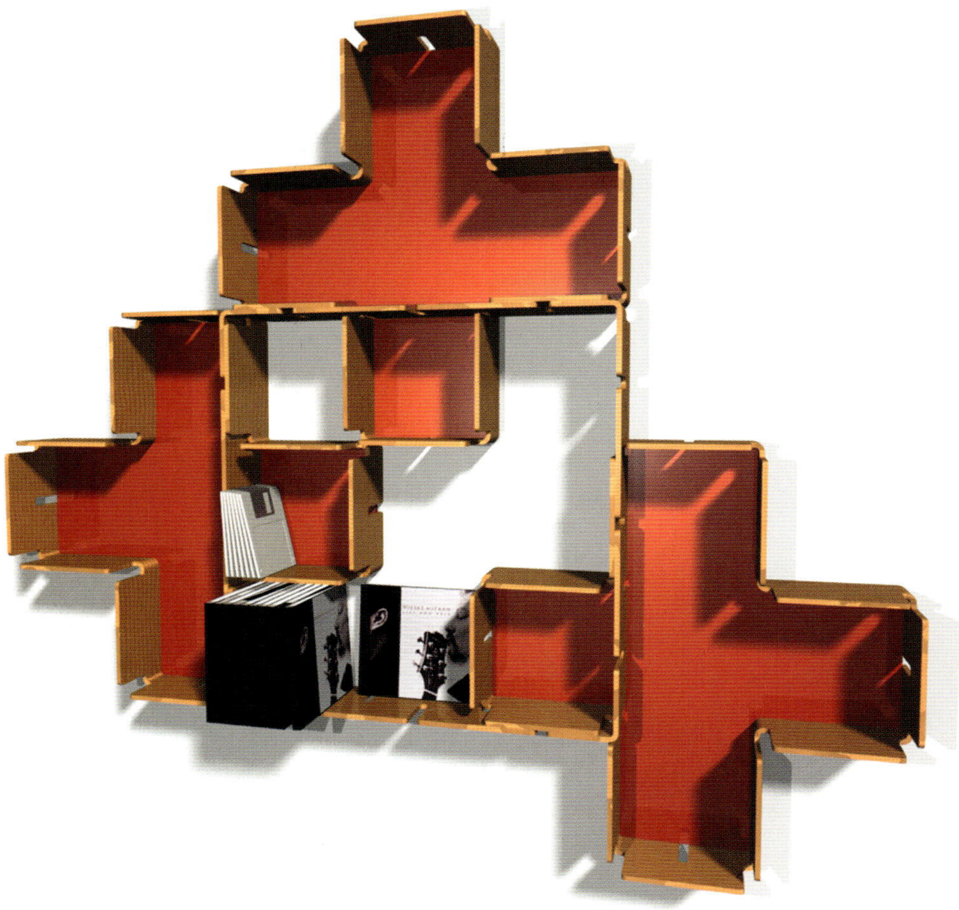

Title

MAGNETISCH /
MAGNETIC TALBLE

Dimension

160 X 80 X 76 CM,
190 X 80 X 76 CM

Design / Photo / Client

JÖRG GÄTJENS
DESIGN DEVELOPMENT

Quantity

PROTOTYPE

Awards

BEST OF
SALONESATELLITE,
MILAN 2004

Year

2004

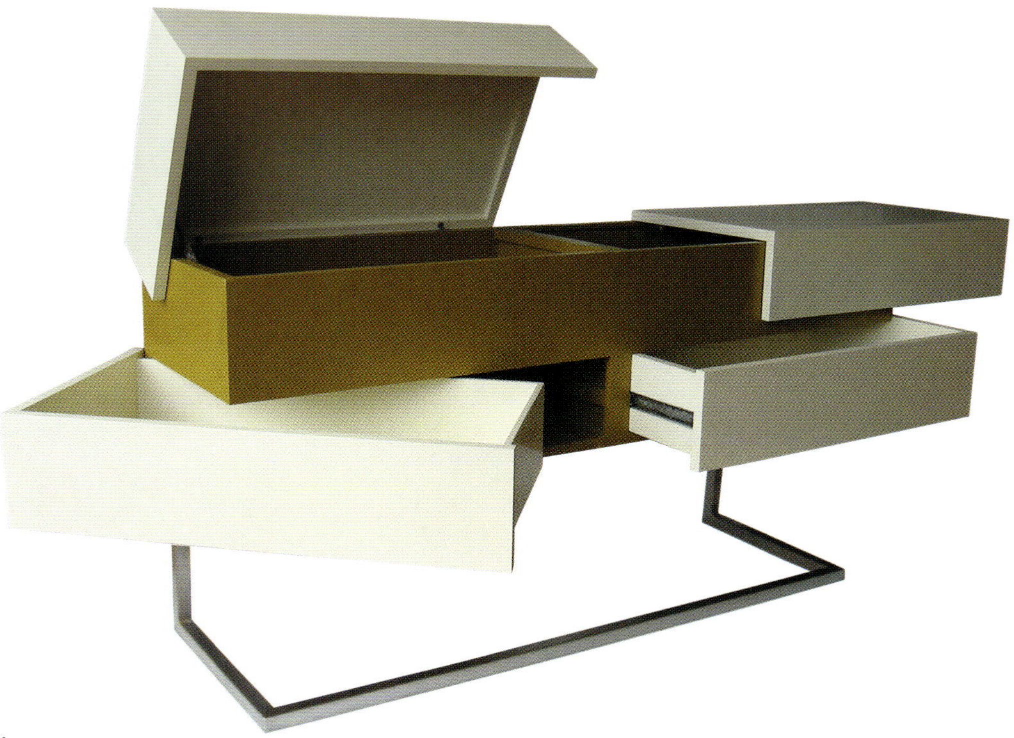

1

2

Title

EXPECT CABINET 1
REBELLIOUS CABINET I 2

Dimension

130 X 70 X 40 CM 1
100 X 120 X 40 CM 2

Design / Manufacture

OOOMS

Quantity

PROTOTYPE

Year

2000 1
2005 2

> See also:
Page 090-091, 101, 134, 152, 234

Title
"GREEN DAY" SERIES

Dimension
VARIOUS

Design / Client
PD DESIGN STUDIO

Photo
MR. DAIJIROU OKADA

Manufacture
SELKON TECHNOS CO., LTD.

Quantity
PRODUCE ON REQUEST

Collection
THE SAN FRANCISCO CHAPTER
OF THE AMERICAN INSTITUTE
OF ARCHITECTS,
THE CENTER FOR ARCHITECTURE
+ DESIGN GALLERY
(FROM NOV 2–DEC 31, 2006),
HAAZ, TURKEY

Year
2005

> See also:
Page 232, 237

Title
SIZE? (I.E.: NOAILLES, BELT, SAMPLE) CARPET
Dimension
BELT : 25 X 120 CM
NOAILLES : (CHAIN) 80 X 45 CM
SAMPLE : 220 CM DIAMETER
Design
ADRIEN ROVERO STUDIO
Photo
PHILIPPE JARRIGEON
Client
GALERIE KREO, ADRIEN ROVERO
Quantity
30, LIMITED EDITIONS
Awards
JURY PRIZE, VILLA NOAILLES FOR DESIGN PARADE, 2006
Year
2006

> See also:
Page 042-043, 045, 088, 096-097, 102, 239

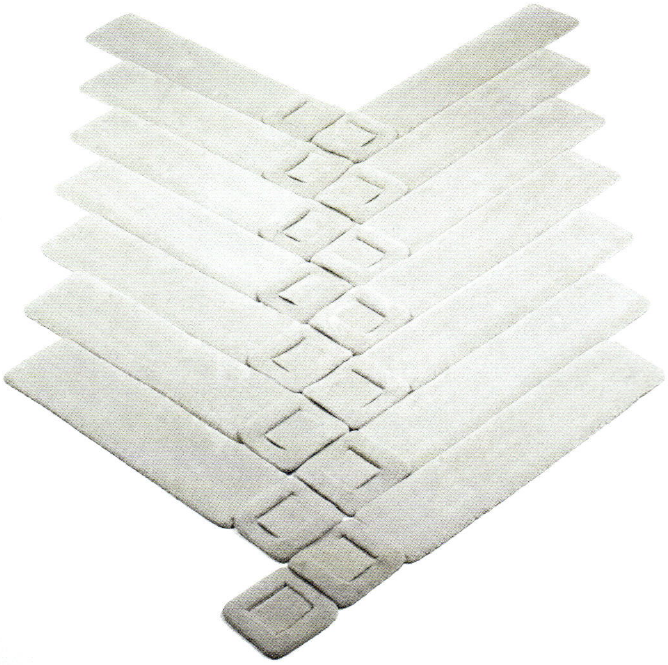

Title
ZIPZIP
Dimension
75 X 75 CM
Design
STERNFORM
PRODUKTGESTALTUNG
Photo
STERNFORM
Manufacture
PLING-COLLECTION,
GERMANY
Year
2006

> See also:
Page 010-011, 028-029,
034-037, 216-217, 241

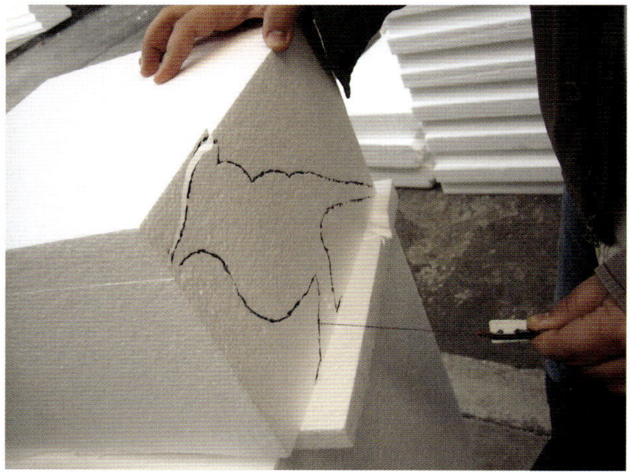

Title
FILLING THE CAT
Dimension
80 X 40 X 30 CM
Design / Client
NEL
Photo
EMILIANO GODOY / NEL
Manufacture
FRIGOCEL MEXICANA
Year
2006

> See also:
Page 243-245

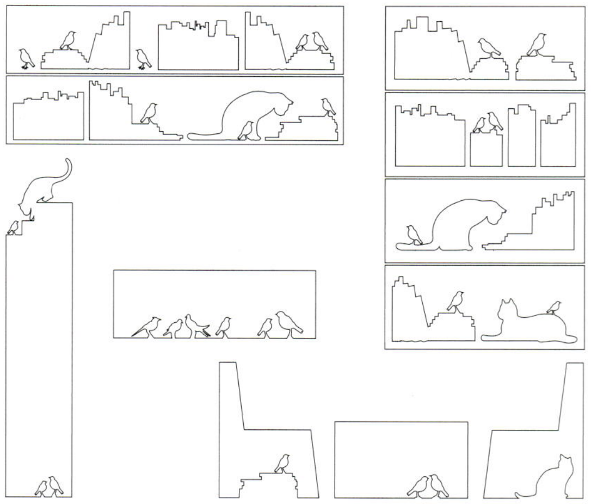

L U N A

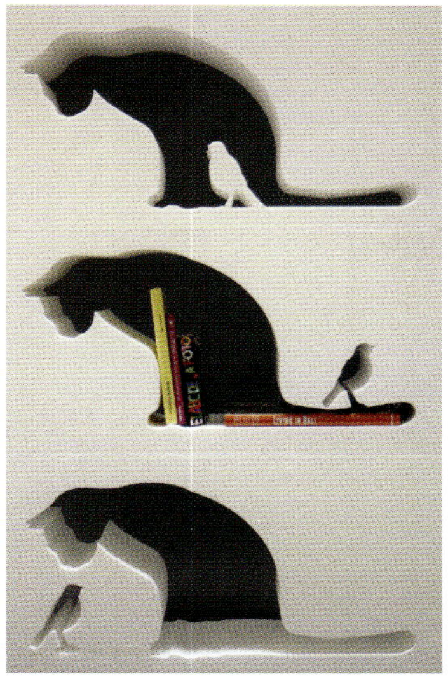

Title
IVY
Dimension
7 DIFFERENT ELEMENTS
AS MODULAR SYSTEM,
APPROX. 75 X 40 X 30 CM
Design / Client
JORIS LAARMAN STUDIO
Photo
ANITA STAR
Manufacture
AXIS, ROTTERDAM
Collection
DSM ART COLLECTION
Year
2003

> See also:
Page 052-053, 247

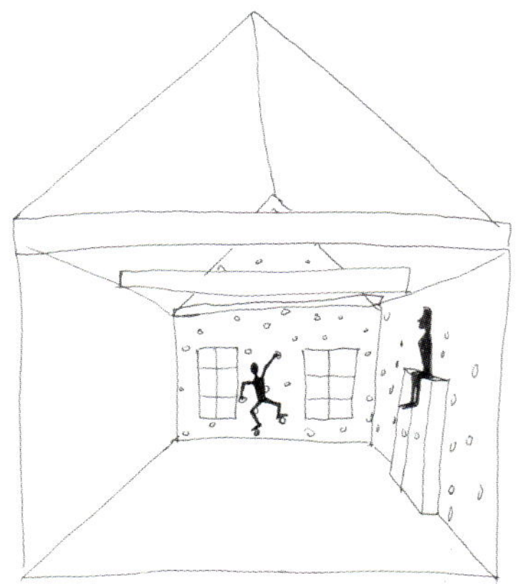

Title
CERAMIC CACTUSES
Dimension
APPROX.
1000 X 90 X 70 CM
Design / Client
HAYONSTUDIO
Photo
HAYONSTUDIO,
NIENKE KLUNDER
Manufacture
BOSA

Quantity
7 PER COLOR,
LIMITED EDITIONS
Collection
MEDITERRANEAN
DIGITAL BAROQUE
Year
2003

BIOGRAPHY

&DESIGN / :: DAN PEARLMAN / ADRIEN ROVERO STUDIO / AFRODITI KRASSA LTD / ALICE WANG / ANDRÉ KLAUSER / AUTOBAN / BKM DESIGN WORKING GROUP / BLEACH DESIGN, LDA / BURO VORMKRIJGERS / CHRIS KABEL DESIGNS / COMMITTEE / CONCEPTS AND IDEAS FOR COMMERCIAL PURPOSES S.L. / DAMIAN O'SULLIVAN DESIGN / DANESE S.R.L. / DESIGNFENZIDER / DIALOG05 / DOMINIC WILCOX / DRESSING / EL ULTIMO GRITO / ELI-GUTIERREZ STUDIO / ERICH GINDER STUDIO / FEINSTET / FIFTY-EIGHT B / FORM US WITH LOVE® / GLITNER / HAYONSTUDIO / HELMUTSMITS.NL / JASON MILLER STUDIO / JEREMY WALTON / JOE MALIA / JÖRG GÄTJENS DESIGN DEVELOPMENT / JORIS LAARMAN STUDIO / JUDITH SENG / JULIAN APPELIUS - BUREAU FOR PRODUCT DESIGN / KARIM RASHID INC. / KARIN VAN LIESHOUT / KYOUEI DESIGN / LA TÊTE AU CUBE / MAARTEN BAAS / MATTHEW KROEKER INDUSTRIAL DESIGN / MICHAEL SANS BERLIN / MIO / MISS GESCHICK & LADY LAPSUS / MIXKO / MOOOI / MULTI / NADINE JARVIS / NEL / OLSCHEWSKI PRODUKTDESIGN / ONTWERPWERK MULTIDISCIPLINARY DESIGN / OOOMS / PARK PLUS / PD DESIGN STUDIO / PROPAGANDIST COMPANY LIMITED / PURE NOMADE / QUBUS DESIGN STUDIO / RAFZ DESIGN STUDIO / REDDISH / SEEDZ LTD. / SHAY ALKALAY / SHOOT THE STYLIST! / STAUFFACHERBENZ / STEFANO CASSETTI DESIGN OFFICE / STERNFORM PRODUKTGESTALTUNG / STUART HAYGARTH / STUDIO VAN EIJK & VAN DER LUBBE / STUDIOMOLD / SYLVAIN WILLENZ DESIGN STUDIO / T.N.A. DESIGN STUDIO / THE DESIGN CAN / THEO HUMPHRIES (3EYES) AND PHILIP WORTHINGTON / TIM PARSONS / TOBIAS WONG / VITAMIN / VLAEMSCH ( ) / VLIEGER & VANDAM / YAEL MER

&DESIGN

A design team formed by four product/interface designers in 2005. Members: Tetsu Miyazawa, Shigenori Ichimura, Keiichi Minamide, Maho Kusano.

Page 113

:: DAN PEARLMAN

:: dan pearlman stands for design, quality and experience. The name of the agency acts as an umbrella for the two companies, dan pearlman brand architecture and dan pearlman experience architecture. This strategic arrangement allows the brand to become an experience and the experience to become a brand. The synergies thus generated benefit all dan Pearlman projects. Design is not considered in isolation but placed within an overall context.

Page 230

ADRIEN ROVERO STUDIO

Born in 1981 in Pompaples (Swiss), Adrien Rovero has been assistant and teaching in the Ecole Cantonale d'Art de lausanne (ECAL) since his Master diploma in industrial design in 2006 there. He opens his studio in the same year and took part in numerous exhibitions. He has worked with producers such as DIM (die imaginäre manufaktur) in Berlin, Tisca Tiara Carpets in Bühler (Swiss), Gallery Kreo in Paris and Suck UK in London. Mentionned in the international press, Adrien Rovero has been awarded a number of prizes including the D&AD Nesta product Innovation Awards for 2003 and 2004 and the jury prize at the Villa Noailles in France.

Page 042-043, 045, 088, 096-097, 102, 238-239

AFRODITI KRASSA LTD

First graduated in product design at Central Saint Martins College, London in 1996, Afroditi Krassa (born London, 1974) was then employed by Seymour Powell Ltd. and Priestman Goode. In 1999, she again studied product design under Professor Ron Arad at the Royal College of Art, London. In 2002, she founded award-winning design consultancy Afroditikrassa Ltd. which has soon built an impressive international client list including Pret a Manger, Topshop and John Lewis in London, E&Y and Cassina in Japan, WMF and Vapiano in Germany, etc. She was recently appointed Creative Director of Itsu Ltd, London as well as Associate Lecturer at Central Saint Martins College, London.

Page 068-069, 169, 200-201, 207

ALICE WANG

Alice Wang, already the First Class Honours degree holder in product design of Central St. Martins College of Art and Design, London in 2006, is currently taking her Master diploma in interaction design at the Royal College of Art, London. During her studies, she took part in numerous design shows in United Kingdom and Japan, such as the New Designers show in London and Designers Block Tokyo in Japan.

Page 162-165

ANDRÉ KLAUSER

German born André Klauser set up his design studio in London in 2002 after finishing his MA at the Royal College of Art. In his design he often draws inspiration from the banal and everyday. His work has been internationally exhibited and published. He works for a number of international clients including Thorsten van Elten (UK), Authentics (Germany) and Adal (Japan).

Page 170-171

AUTOBAN

Founded in 2003 by architect Seyhan Ozdemir and interior designer Sefer Caglar, Autoban is a combination of works on architecture, interior and product design. Every single Autoban product has a certain character, reflecting the designers' quirky approach to 'almost cliché' items. Starting point for each of these products is the material itself and its capabilities. As their names imply, the range consists of furniture and lighting units that are inspired by the nature and the rituals of daily life.

Photo: Emre Dogru/20ML
Page 146-149

BKM DESIGN WORKING GROUP

In 2004 the successful collaboration of bkm design working group began. Three different backgrounds in industrial arts, three courses of study in product design in Vienna and Milan and various experiences in product exhibition and experimental design, they are Katharina Maria Bruckner, Herbert Klamminger and Stefan Moritsch. They believe that only their rigorous integration can truly lead to innovative products. bkm attributes the quality of its designs and distinctiveness of its works to this combination of penetrating insight, artistic integrity and the indefatigable search for form.

Photo: Roland Ferrigato
Page 032-033, 076-077

BLEACH DESIGN, LDA

Bleach Design believes that new concepts appear every day, new ideas that urge reinventing themselves, day-by-day at the same speed the world goes by. It is a quick world, either in accepting than in refusing; what is good today can be bad tomorrow; what is in limelight today will be forgotten tomorrow. So, we aren't looking for a place in history, we are just trying to produce whatever can make a difference in every days life, either through the sensitivity of the customer than in the memories of the ones who produce them. We don't want to work 'for' but to work 'with', in a continuous project that never ends, but always renewing itself.

Page 132

BURO VORMKRIJGERS

Buro Vormkrijgers (Bvks) is a Dutch design studio founded by Sander Mulder [1978] and Dave Keune [1978], both graduated at the Design Academy in Eindhoven, the Netherlands. Over the years, Bvks has become an international operating studio which specializes in the field of professional furniture, lighting and interior design. It combines work for international business and private customers with concepts developed under their own management. The results have been published worldwide in leading magazines and TV programs, and have been purchased to leave their traces in homes, offices, galleries and museums all over the world.

Photo: Raoul Kramer
Page 044, 120, 122-123, 135, 194-195, 218-219

CHRIS KABEL DESIGNS

Now based in Rotterdam, Chris Kabel has studied industrial design at the Design Academy of Eindhoven, where he was noticed by Droog Design. His diploma project Sticky Lamp was included in their catalogue in 2002, the beginning of an ongoing collaboration including commissions and exhibitions. Mooi and Royal VKB have followed, and some of his products have entered the collections of the Boijmans van Beuningen, the stedelijk museum amsterdam and the FNAC. Nominated in 2003 for the Rotterdam Design Prize with his 1totree lamp, he has had a solo show in Paris in 2006, another sign of his recognition as a promising independent designer.

Page 058-059, 062-063, 084-085, 182-183

COMMITTEE

Committee is particularly interested in the way aesthetics and visual culture impact on our lives. They are curious about the detailed nature of people's relationship to such languages and mainly work on furniture, lighting, textiles, wallpaper and interiors. It is run by Clare Page and Harry Richardson, both born in 1975. They met whilst studying Fine Art at Liverpool Art School and subsequently studied Machine Knitting and Hand Crafted Cabinet Making respectively. They then developed the 'Gallop' workshop and gallery space, in a previously derelict shop in Deptford, SE London and formed the Committee design studio in 2001.

Page 139

CONCEPTS AND IDEAS FOR COMMERCIAL PURPOSES S.L.

Formed in Barcelona and Milan as an interior and industrial specialist, Martí Guixé (1964) is periodically working as a design advisor in Seoul in 1994, he formulated a new way to understand the culture of products. In 1997, he started to exhibit his work that characterizes on the search for new product systems, the introduction of design in food ambits and presentation through performance. His non-conventional gaze provides brilliant and simple ideas of a curious seriousness. He is based in Barcelona and Berlin and work as designer for companies like Authentics, B-Sign, Camper, Cha-cha, Chupa-Chups, Desigual, Droog Design, Esencial mediterraneo, Imaginarium, Isee2, Nani Marquina, Very Lustre, Saporiti.

Page 136-137, 186

DAMIAN O'SULLIVAN DESIGN

Damian O'Sullivan was born in England in 1969 of Franco-Irish parents, he grew up in the Netherlands where he now runs his own design agency, Damian O'Sullivan Design. Founded with a mission to provide 'fresh and original work that reaches beyond the pale and the predictable', its initial assignments have been as diverse as they have been interesting: an 18m high landmark barometer for London's Holland Park roundabout, the development of a water sculpture for Anish Kapoor, external OnStream back-up drives, a perfume bottle for Comme des Garçons, a home cinema system for Pioneer, concept development for Lexus Cars Europe, shoe proposals for Paul Smith, Asics and Camper.

Page 106-109

DANESE S.R.L.

DANESE was instrumental in defining the DNA of Italian design. Since 1957, the year of its founding, Danese's evolution has been dynamic and strongly innovative, a perfect fusion of design culture, creative research and production technique within not only a conceptual framework but also a concretely productive one expressed through simple objects of great quality. Danese's production embraces a range of archetypal objects whose age and style are hard to define. With a catalogue that points in all directions, the whole Danese collection is composed of versatile objects divided into three main themes — living, working, lighting — a modern answer to the relationship between people and their space.

Page 129

DESIGNFENZIDER

Ron Gilad is a philosopher of the everyday object. Israeli by origin, Gilad studied industrial design at the Bezalel Academy of Arts and Design in Jerusalem. He taught conceptual 3D design at the Shenkar School of Engineering and Design in Israel. In 2001 Gilad relocated to New York and co-founded Designfenzider, a test lab for his ideas. His hybrid objects combine material wit with aesthetic play; they sit on the fat, delicious line between the abstract and the functional. It asks unceasing questions in 3D form, and fabricates answers that create an arena for fertile doubt.

Photo: Allison Bradley
Page 012-013

DIALOG05

dialog05 was co-founded by Christian Busse, Roman Gebhard, Markus Johannes Mayer, Jochen Scholer and Ina Jade Seng in May 2005. All founding members met working as industrial designers at a design consultancy in munich and participated in award-winning design programs for clients such as Sony, Siemens, HP, Motorola, Procter & Gamble, LG Electronics, Whirlpool etc.

The aim of dialog05 is to explore design beyond product development for the industry. The first group exhibition was on show in the gallery 'die erste reihe' in munich from December 2005 to January 2006.

Page 092-093

DOMINIC WILCOX

After graduating from the Edinburgh College of Art, Dominic Wilcox moved to Japan for a year before returning to the UK to study an MA at the Royal College of Art. During the 2-year period of intensive critical debate and creative experimentation he developed a series of objects including War Bowl, Honesty Stamps and Glove. After leaving the RCA in 2002, he started a collaboration with fellow RCA graduate Steve Mosley called Mosley meets Wilcox. One major MMW project was a collaboration with Rock photographer Mick Rock which ended in a collection of objects exhibited in Sept 2004. In 2006 Dominic went solo, and is currently developing new work, experiments and ideas.

Page 130-131

DRESSING

The three designers Veronika Becker, Heike Scheller and Judith Seng met during their studies of product and process design at the University of the Arts Berlin. They occasionally collaborate on projects such as a clothing collection for furniture called 'Pick Pocket, Cosy Cover, Big Socks' which was presented it in various exhibitions.

Page 078-081

EL ULTIMO GRITO

El Ultimo Grito is the creative partnership of Rosario Hurtado and Roberto Feo, who studied Industrial Design at Kingston University and the Royal College of Art in London respectively. Founded in 1997, their studio is now based in London. Their work continuously researches human relationships with objects and culture, exploring them across disciplines in a wide variety of projects for clients such as Magis, Mathmos, Nola, Marks&Spencer and British Airways. Their work has been widely published, awarded and exhibited, being part of permanent collections such as the Stedelijk Museum, Victoria&Albert Museum, Manchester Art Galleries or the Kulturhusset.

Page 202-203

ELI-GUTIERREZ STUDIO

Eli Gutierrez was born in Valencia in 1980. After graduating from the ESDI CEU of Valencia, Spain, she moves to London to study in MA product design at The Royal College of Art, London. Her work has been exhibited in numberous exhibitions like the NUDE Saloon, the MUVIM Museum in Valencia, DE+Modoloco, Livingspace, Made by Machine in the Furniture fair of Milan. She is recently taking part in the Disturbance Exhibition at the Great Eastern Hotel in London too. Currently she is working for differents Spanish companies as freelance.

Page 210

ERICH GINDER STUDIO

Erich Ginder Studio was founded in the fall of 2004 with the conception of his ghost collection, a series of objects created entirely by hand from solid cast resin. Before focusing on a career in design, he studied at Chicago's School of the Art Institute and worked as a production artist in advertising. Now based out of Seattle, his studio focuses on well-crafted objects that attempt to blur the line between art and design.

Page 048-049

FEINSTET

The collective Wiedenmann & Averdiek analyse the literal sence of things in a playful way and not their function as they are defined in dictionaries. Their focus is on the intended activity. And the search for a simple solution then forms the object.

The two designers from Berlin met in 1999 during their studies at the University of the Arts Berlin and since then have been working collaborativly to improve the world.

Page 231

FIFTY-EIGHT B

An award winning furniture design, project management consultancy studio, Fifty-Eight b designs, develops and project manages collection and bespoke furniture and homewares product for industry, retailers, architects, manufacturers and brands.

Under their Gallery Edition range, Fifty-Eight b also produce limited edition, quirkier, maker-made art and design pieces which are fabricated to order.

Integral to studio practice is the vision of designing and offering considered, inspired and responsible interior product for the contemporary furniture and accessories design market.

Page 046-047, 160-161

FORM US WITH LOVE®

FORM US WITH LOVE® is a young and growing, global design company set up in Sweden in 2005. Our focus is on design within the segments of products and furniture. We work unconventionally and we let every product develop according to its own specific background variables. We are inspired by the interaction between people and products, driven by the genuine love for design and creating every product through a unique design process. Our projects range from lighting design for Design House Stockholm, Uno design and Zero to furniture design for Materia, Mitab and Wilo, to watch and jewellery design for Axcent of Scandinavia.

Page 188, 224-229

GLITNER

Tobias Mellberg was born on 12 May 1969 in Vaasa, Finland. He studied Sculpture at the Nykarleby Art School in 1993-1997 and Goldsmithing and Design at Lahti Polytechnic in 1997-2001. In 2002, he formed the jewellery studio Glitner in his home town.

Page 103

HAYONSTUDIO

Jaime Hayon was born in Madrid, Spain in 1974 and trained as an industrial designer in Madrid and Paris. Began working as a researcher in Fabrica, Italy in 1997, he began his individual career in 2004 and has kept his hands full with eclectic projects. From toys, to furniture and interior design as well as artistic installations, his boldness has been transcending the borders of the often-separated worlds of art and design, merging his own style with ease between the two. Currently with a new base in Barcelona, Hayón has been awarded by Wallpaper as one of the 10 breakthrough creators worldwide as well as the 2006 Elle Deco International Award and the Icon Magazine Best Show award for the 2006 London design week.

Page 248

HELMUTSMITS.NL

Helmut Smits (1974) is a multidisciplinary visual artist based in Rotterdam, The Netherlands.

Page 190-191

JASON MILLER STUDIO

Jason was born in New York in 1971. He received a BA from Indiana University in 1993 and an MFA from New York Academy of Art in 1995. Between 1995 and 2001, Jason worked as a studio assistant for Jeff Koons, an art director for Ogilvy and Mather and a designer for Karim Rashid. In 2001, Jason began to work independently and designs everything from furniture to tchotchky to interiors. He has been featured in numerous publications including Metropolis, The New York Times, Thirty4, etc. Jason has received several awards including the Bombay Sapphire 'Rising Star Award'. Jason has developed products for companies including Areaware, Benza, The Conduit Group, etc.

Page 064-065

JEREMY WALTON

Born in 1972, British Product Designer Jeremy Walton graduated from Central Saint Martins College of Art and Design, London in 1996. After graduation he worked for a spell in Asia, subsequently travelling through China and Russia on the Trans-Siberian Express, before settling in Denmark in 1998.

Walton's inventive and original work has been featured in exhibitions both nationally and internationally. His Walton.dk products is inspired from his experience of life in Denmark. Although his work draws from his immediate locality, this does not stop it from gaining international sales and interest, most notably the MoMA New York retails stores.

Page 089

JOE MALIA

Joe Malia works at the intersection of graphic design, interaction design and product development. He has worked with professionals from a variety of backgrounds including, writers, curators, computer programmers, sociologists and textile designers.

Page 178-179

JÖRG GÄTJENS DESIGN DEVELOPMENT

Jörg Gätjens was born in Duisburg in 1965. He studied industrial design in Hannover and after graduation, he worked for Nils Holger Moormann, Aschau, in the team of design management. In 2002 he opens his studio in Cologne. His work is exhibited internationally such as Designparcours in Munich, Talents, Tendence Lifestyle in Frankfurt and Salonesatellite, Salone del Mobile in Milan.

Page 233

JORIS LAARMAN STUDIO

Joris Laarman was born in Borculo, the Netherlands in 1979. In 1997 he began his study at the Design Academy Eindhoven to combine architectural design and fine arts. In 2003, he started his own independent Joris Laarman Laboratory, which is a studio for innovation, aesthetics, curiosity, freedom and more. It concentrates on architectural interventions and product design with a poetic kind of engineering. Next to a collection of highly unique products JL works for several well-known international design companies and galleries. He recently contributed in articles and seminars for Domus magazine and he was a guest teacher at European universities as well as the Architectural Association (AA) and Design Academy Eindhoven.

Page 052-053, 246-247

JUDITH SENG

Since 2003 Judith Seng works on self-initiated and commissioned projects for different studios and clients spanning the field of cultural research to design. According to project requirements she collaborates with a team of different experts.

She studied Product and Process development (Hans Nick Roericht, Enzo Mari) at the University of the Arts in Berlin. After graduating in 2001, she stayed as assistant teacher of Achim Heine, Volker Albus and Christophe Marchand. Establishing her own studio in 2003, she continues occasionally teaching as guest teacher.

Page 118

JULIAN APPELIUS – BUREAU FOR PRODUCT DESIGN

Founded in 2003 in Berlin, the Bureau for Product Design focuses on conceptual products, products which play with our learned habits of cognition and behaviour. Founding principles of formal reduced design include challenging us to re-interpret our context while realizing the full potential of material and technical design aspects.

With the lamp Left or Right, the first marketable product was developed, produced and brought to market through independent organization of a distribution channel.

Additional activities include graphic, interior, and event design.

Page 196-197, 212-213

KARIM RASHID INC.

Karim Rashid is a leading figure in the fields of product and interior design, fashion, furniture, lighting and art. Born in Cairo, half Egyptian/ half English, and raised in Canada, Karim now practices in New York. He is best known for bringing his democratic design sensibility to a mass audience. Working with an impressive array of clients, Karim is radically changing the aesthetics of product design, and the very nature of the consumer culture in which we now live. To date he has had some 2000 objects put into production. In addition Karim has successfully entered the realm of architecture and interiors.

Page 022-023

KARIN VAN LIESHOUT

As a product designer, Karin van Lieshout is not as much focused on the technological progress, as on the meaning of things, their cultural and emotional value. Inspired by handcraft, combined with new materials and thoughts, her goal is to develop new products with a visual elegance.

Page 054-055, 090-091, 184-185

KYOUEI DESIGN

Kouichi Okamoto was born in Shizuoka City, Japan in 1970. Other than music, three works (BalloonLamp, Color Light, Random) were announced as new developments in 2003. Kyouei design was established the next year. In 2006, he took place in the 'TWO DAYS & FOUR PRODUCTS' exhibition at Shizuoka Prefectural Museum of Art. He also participated in various design events like ICFF(NY) in May, the 100% Design Tokyo and Saint etienne 5th biennial International Design Festival in November.

Page 187, 192-193

LA TÊTE AU CUBE

Established in Paris, France in 2005 by Jerome Fischbach and Thierry Galloni d'Istria, La tête au cube's objective is to initiate, create and develop new products (with the exception of clothing) destined for the consumer market, for the home and 'for the individual'. The company name has apparently something to do both with anatomy and with mathematics, the company's slogan 'gentiment décalé', in French, is almost impossible to translate into English. The approach is based on a solid marketing rhetoric and addresses both niche and mass distribution markets. They aim to collaborate with young creatives and designers by offering an opening to those who otherwise might not have had the possibility to address larger markets.

Page 104-105

MAARTEN BAAS

Dutch designer Maarten Baas was born in Arnsberg, Germany in 1978 and from 1979 grew up in Burgh-Haamstede and Hemmen in the south and centre part of the Netherlands. In 1996 he studied at the Design Academy and still in school his first design, candleholder 'Knuckle', was taken in production bij Pol's Potten.

In 2005 he started his collaboration with Bas den Herder to form, who became responsible for the production of all pieces. The foundation of studio Baas & den Herder made it possible to produce Maartens unique pieces on a larger scale, but yet all hand made in Holland.

Page 154-155

MATTHEW KROEKER INDUSTRIAL DESIGN

Matthew Kroeker is a freelance industrial designer developing a range of products including furniture, housewares, accessories and consumer products. Other than creating objects for major manufacturers, he produces experimental and forward-thinking furniture prototypes which have been exhibited in Toronto, New York and Milan. Aiming to enhance the rituals of daily life, the scope of his work has become increasingly diverse ranging from interiors for urban transit buses, medical devices, housewares, and upholstery which have allowed him to explore a blend of new and unconventional materials while offering a strong balance between concept, form, and efficient manufacturing.

Page 211

MICHAEL SANS BERLIN

Based in Berlin, Germany, Michael Sans is working for a wide range of clients and industries, including Automotive, Fashion, Furniture and Lighting as well as Household-Appliences. He has been appointed guest professor at the 'Hochschule für Gestaltung Braunschweig' where he is teaching in the department of industrial design. He is trained as a cabinet maker, received a bachelor degree from the Art Center College of Design in Pasadena, USA and has worked as Automotive Designer for Volkswagen AG, Germany. Before opening his studio in 2003 he finished the Royal College of Art in London with a Master Degree.

Photo: Frank Peters, Berlin
Page 100, 124-125

MIO

MIO was founded by the Salm brothers (Isaac and Jaime) in 2001 with the objective of combining business rigor with environmentally and socially progressive design. Based out of Philadelphia, MIO is a design company dedicated to creating the most sustainable and innovative design experiences in the market. Guided by the company's philosophy of 'Responsible Desire' MIO has introduced design that strategically harnesses the most eco-intelligent aspects of conventional manufacturing, marketing and distribution methods. This approach has resulted in the integration of existing technologies and industries into a profitable value-added system that is both socially and environmentally responsible.

Page 206, 220

MISS GESCHICK & LADY LAPSUS

Miss Geschick & Lady Lapsus are founded by Olga Bielawska (born in Warschau, 1979) and Astrid Schildkopf (born in Nürnberg, 1980) in May 2006. Both of them have been studying product design at the Bauhaus-University in Weimar since 2001. Their products are dealing with faux pas or delicate themes of the everyday life which they don't want to eliminate but put them in a positive context. That way small mishaps and such will become a stylish and communicative element in life. They believe that design is essential for our everyday life. Especially with their concept, they try to improve that by being funny and effective at the same time. It is important for them to create new perceptions by playing with different perspectives.

Page 038-039

MIXKO

MIXKO is design partnership formed by Goldsmiths College graduates Nahoko Koyama and Alex Garnett. Both had their work selected by the British Council during Milan Furniture Fair 2003, and featured in influential magazines whilst undergraduates. Since then they have continued to earn critical acclaim for their creations at various international design events.

MIXKO aim is to create items that have a wide appeal, inspire happiness and possess a satisfying simplicity. Nahoko and Alex are continually seeking fresh design ideas to meet various everyday needs with original thinking.

Page 020, 026-027, 094-095, 153, 166-168

MOOOI

Moooi first presented her innovative collection, under creative direction of Marcel Wanders, at the Salone del Mobile in Milano, April 2001. Since then, Moooi has started cooperations with internationally well-known creative people like, Ross Lovegrove, Li Edelkoort, Erwin Olaf, Joep van Lieshout and Jurgen Bey. Moooi leads the understanding of culture, human needs and technology, contributes by creating the most successful innovative design-collection ever and creates unprecedented value and opportunity for its customers, employees, investors and partners. This is their challenge and vision, which need our help to realise it and get hold of it.

Photo: Erwin Olaf
Page 138

MULTI

Kathleen Hills set up her design studio Multi in 2002 after graduating with a 1st class honours degree from Central St Martins and an MA in Ceramics and Glass from The Royal College of Art.

Kathleen's work has been published in various national and international magazines and newspapers including The Telegraph, The Observer, Blueprint and Elle Decoration. This year she was selected in UK Marie Claire's hot 100 List to represent design.

Multi's products are now sold in major retail outlets such as Conran UK and Paris, Libertys, Heals and Le Bon Marche. Multi's celebrity client list includes Cate Blanchett.

Page 150-151

NADINE JARVIS

Born in 1982, Nadine Jarvis is a London based product designer. Since completing her studies at Chelsea College of Art and then Goldsmiths College, she has exhibited in London's Design Museum and her work has received multiple awards, including sponsorship from the British Council. Through her projects she has fortified her reputation as a conscientious and poetic young designer. Her objects reflect her interest in the idiosyncratic, temporality and her experimentation within the zeitgeist of unusual materials and technologies.

Page 172-173, 175

NEL

NEL is a collective of independent designers who join forces to explore, through concept and form, the interaction in design between traditional materials and techniques, and new processes or unusual applications. NEL is founded in the year 2004 in Mexico City by Cecilia Leon de La Barra, Hector Esrawe, Ricardo Casas and Oscar Nuñez. In the year 2005 Emiliano Godoy joins NEL to present the second collection of NEL.

Page 242-245

OLSCHEWSKI PRODUKTDESIGN

David Olschewski was born in Cologne, Germany in 1977. After studying product design at the Academy of Design / ecosign, he founded his own design studio, Olschewski Produktdesign in 2006. In addition to designing events, the team concentrates on conceiving innovative products whose functionality is coupled with high aesthetic impact. The products have an emotional appeal that invites the user to partake in an experience that is both sensual and contemplative. The young team draws its inspiration from first commissions as well as regular participation in exhibitions including the Cologne Passagen (2002 - 07) and the Interieur 06 Biennial in Kortrijk.

Page 074-075

ONTWERPWERK MULTIDISCIPLINARY DESIGN

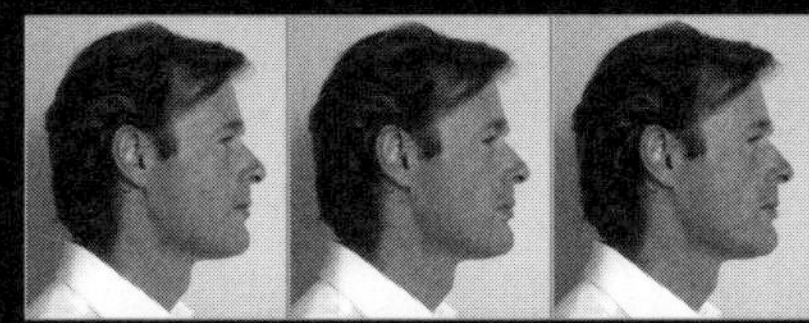

The right combination of function, form and material is what makes a successful product. Ontwerpwerk makes striking consumer products for international top brands. They are known for their creativity and clarity in product concept, design and manufacture. In this process, they make optimum use of the possibilities of new technologies and materials, taking distribution issues and market requirements into account. Several Ontwerpwerk products have won international design awards and have been included in museum collections all over the world. Their 30 designers work as a team in four disciplines: graphics, architecture, product and web design. Ontwerpwerk's quality is that all these specialties are available in-house.

Page 041

OOOMS

OOOMS is the name of a design studio based in Eindhoven, The Netherlands. This studio was founded in 2000 by Guido Ooms. The studio mainly works on self-initiated projects, mostly concerning furniture and other interior products. The main objective for OOOMS is to develop most objects and experiments into fully functional and commercial industrial products, preferably in cooperation with manufacturers.

Page 090-091, 101, 134, 152, 234-235

PARK PLUS

Zinoo Park is that rare kind of product designer that strikes a nice balance between the functional and the emotional. Hailing from the Royal College of Art in London, Zinoo Park is now working as a creative director for design studio Park Plus, art director for the music videos and an artist manipulating Coca-cola bottles. His works have been shown in Seoul, as well as in Milan, London and at design exhibitions across Europe. One of his most prominent works is the erotic dishes series — a series of functional dishes for eating from that slightly hint at the other most basic of human needs. Also, Zinoo Park's ongoing personal Coca-Cola series project is new take on the omnipresent brand.

Page 066

PD DESIGN STUDIO

pd DESIGN STUDIO was founded by product designer Hideo Hashimoto (born in Kyoto-City, 1968) and Izumi Hamada (also born in Kobe-City, 1972). They specialize in furniture design, indusutrial design,graphic design etc. They aim to reconstruct the significance of existence of things from all viewpoints. To make a new route to their design direction, they formed another design firm, The Design DUO in 2006, which aims at carrying out expression dispatch between people.

Page 232, 236-237

PROPAGANDIST COMPANY LIMITED

Propagandist Company Limited was established in 1994 by a group of advertising experts who made commercials for major consumer products. The group consisted of Satit Kalawantavanich, Boonkiat Korsanan, Yodphet Sudsawad, and Metta Sudsawad. Propagandist tries to capture the innate sense of Thai playfulness, a spirit of fun and the sense of humor that can be found in every aspect of Thai culture around the working environment into products under the brand Propaganda. Since 1996, it acquires plenty of world-class awards such as Form 2001, Red Dot award 2002 and Good Design 2005. Products under the brand are now available in more than 30 countries throughout the US, Europe, Hong Kong and Japan.

Page 031

PURE NOMADE

Pure Nomade is a partnership between two freelance Lilian Adler and Søren Berg, both studied in Design School of Denmark. Their clients include Lego, DSB and BodySoul; and Kompan, Duba and Softline respectively.

Page 204-205

QUBUS DESIGN STUDIO

Qubus design studio was established in Prague in 2002, producing product design, mostly from traditional Czech materials such as porcelain and glass. It designs and produces its own products, Qubus represents lot of Czech designers such as Jiri Pelcl, Anna Torfs, Daniel Pirsc and others. Main designers acting in Qubus are Jakub Berdych and Maxim Velčovský. Maxim Velčovský is now taken in as one of the key person in the new wave of Czech design. Their other activities are graphic design and photography. And their great and well-known collection is called the White collection by Maxim Velčovský.

Page 014-019, 021, 030

RAFZ DESIGN STUDIO

Born in Catania, Italy in 1974, Raffaele Iannello studied Industrial Design at the Polytechnic University of Milan and founded Rafz Design Studio. Since 2002 he designs furniture and industrial design objects for renowned companies such as Flou, Slamp, Viceversa. From 1996 to 2001 he worked as an art director with the major Italian new media agencies, with which he won numerous awards and received international recognitions for creating, among others, the official websites of Bulgari, Chicco and Pioneer. Since 1999, he has been writing on design, technology and contemporary art for different international magazines.

Page 133

REDDISH

Reddish was founded in 2002 by industrial designers Naama Steinbock and Idan Friedman. Both born in 1975, graduated with distinction from the Holon Academic Institute of Technology, Israel. Reddish studio focuses on keeping its designs clear and intriguing, and spends most of the time helping objects feel better about themselves.

During the last year, Reddish studio won the 1st. prize of the inspired by cologne 06 — interior innovation award imm cologne, and exhibited at the V&A museum in London.

Page 072-073, 082-083, 189

SEEDZ LTD.

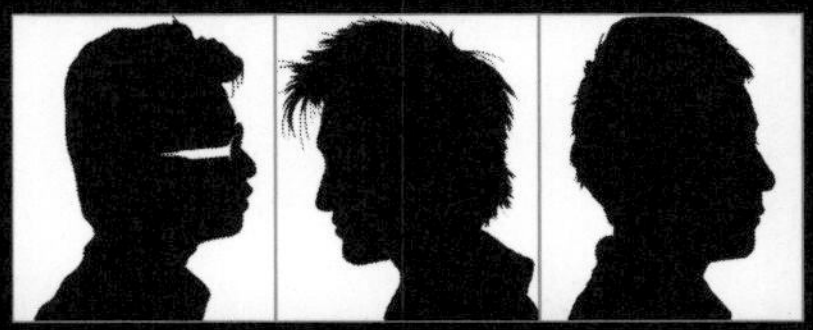

For Seedz Ltd. (formerly called CHILLICHILLY), design is recognizing the art of living. Through their work, they hope to give attention to ideas that are clever and contemporary without being media and trend driven. Though they are constantly submerged in a world that claims to understand what 'lifestyle' is all about, they make no pretensions about being arbiters of how one should style one's life. Instead, they would like to think of themselves more as story tellers, who with each endeavor push to reveal, narrate, as well as discover the connections between art and living.

Page 121

SHAY ALKALAY

In his work, Shay Alkalay develops new typologies that challenge the users' relationship with physical principals such as reaction to gravity, revolving objects and mobility, emphasizes new ways to use products.

He was born in 1976 in Tel-Aviv and complete his MA in Ron Arad's Design Products Department at the Royal College of Art, London.

Page 126-127, 174, 176-177

SHOOT THE STYLIST!

Brit Leissler has studied design at the Koeln International School of Design as well as the California College of Arts in San Francisco. She has worked in communication and interface design before going to London where she currently studies towards an MA in product design at the Royal College of Art, London.

Her work is addressing the alienating nature of modern urban life, her objects prompt the rediscovery of lost basic emotions — referencing tactile, domestic forms but using touch and play to combat the monotony of urban life. Through her humorous and inviting work, Brit Leissler challenges us to rekindle the lost art of happiness.

Page 180-181

STAUFFACHERBENZ

In 2003, originally trained as a primary school teacher, Nicole Benz (1969) founded StauffacherBenz with Stefan Stauffacher (1965), who was originally a biologist in water-protection. It is a product studio based in Zurich, Switzerland. Since then they have been working on a wide range of projects, starting from appliances and furniture over accessories and jewellery up to trade fair stands. StauffacherBenz do commissioned work for their clients as well as they develop their own ideas, from which smart and surprising products or witty conceptual design statements can result.

Page 128, 214-215

STEFANO CASSETTI DESIGN OFFICE

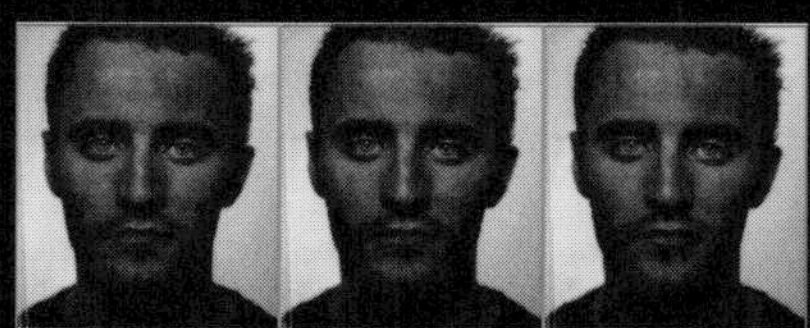

Stefano Cassetti is a Paris based Italian designer with an industrial design high level license from the Politecnico School of Design in Milan. After ENSCI license in Paris, he works in Milan, Garda Lake, Marseille and Paris in product design about travel and interiors.

Page 086-087

STERNFORM PRODUKTGESTALTUNG

Sternform Produktgestaltung by Andrea Grossfuss and Olaf Kiessling, is a young office for product design. They specialize in furniture, accessories and all kind of products surrounding and concerning people. They try to walk through life open-eyed and to react adequately. Additionally, they don't want to think of the functions of things only, but also care about life being as pleasant as possible with them — good handling, a smile, an insight, amazement, a pleasant movement, a lightening.

Page 010-011, 028-029, 034-037, 216-217, 240-241

STUART HAYGARTH

Stuart Haygarth was born in Whalley, UK in 1966. Graduated in 1988 in Graphic Design and Photography at the Exeter College of Art & Design, Devon, UK, he became a freelance photographic illustrator in 1991. Stuart had been a visiting lecturer at Plymouth University Photography Department for ten years since 1995. During all these years, he has participated in numberous design events such as 2005 Designersblock London, part of the London Design Festival, 2005 Installation of Tide chandelier at Comme des Garcons store (Dover street Market) London, 2006 Cassina showroom in Paris and 2006 Cibone gallery in Tokyo, etc.

Photo: Ayumu Gombi
Page 140-145

STUDIO VAN EIJK & VAN DER LUBBE

Both graduated at the Design Academy and the post graduate programme of the Sandberg Institute in Amsterdam, Niels van Eijk and Miriam van der Lubbe started their design studio in 1998, where they work individually but share one studio. Often they collaborate in projects. They focus on product, interior and exhibition design. Work has been exhibited worldwide and purchased by many museums such as Museum Boymans van Beuningen Rotterdam, Centraal Museum Utrecht, Museum voor Moderne Kunst Arnhem, The Dutch Textile Museum Tilburg, Manchester City Art Gallery, WOCEF Korea, Museum FIT New York and Stedelijk Museum Amsterdam.

Page 070-071, 116-117, 158-159

STUDIOMOLD

Studiomold was founded by Brendan Young and Vanessa Battaglia in 2003. The Anglo-Italian design couple started the company with an vision to create the perfect mold for a design studio, one that provides the ideal environment for designers to experiment in, allowing them to achieve their creative best.

Because of their conception, the studio's inventive approach has achieved international recognition as well as receiving numerous design awards. Currently it has successful designs sold by many well-known retailers including John Lewis and Next.

Page 067

SYLVAIN WILLENZ DESIGN STUDIO

Born in Brussels in 1978, Sylvain Willenz studied 3D design in the UK and Design Products at the Royal College of Art in London graduating in 2003. Willenz now works in Brussels on various projects, from product and exhibition design to interior installations. Works are characterized by an eclectic use of materials and processes as well as they are concerned with their relevance in design culture.

Page 060-061

T.N.A. DESIGN STUDIO

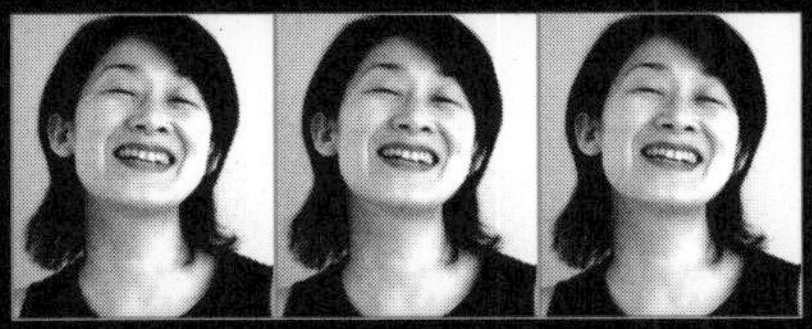

In 1989 Tomoko Azumi (born in Hiroshima, 1966) graduated in Environmental Design at Kyoto City University of Arts, Japan. In 1995 she opened a design studio AZUMI in London, with Shin Azumi and took a master course in Furniture Design at the Royal College of Art, UK. During their partnership, they won numberous awards such as 2000 'Product of the Year' FX International Design Award and 2003 'Best Contribution' 100% Design Blueprint Award, UK. In 2005, Azumi opened t.n.a. Design Studio independently. Recently, she has been a tutoring staff at Royal College of Art and a research fellow at London Metropolitan University.

Photo: Julian Hawkins
Page 221

THE DESIGN CAN

Founded in 2003 by Jeannie Choe and Steven Tomlinson, The Design Can (TDC) is a Brooklyn-based creative team focused on design for exceptional living. Both graduated at Industrial Design program at Pratt Institute in Brooklyn, NY., the two are polar opposites in mind and character, making for an interesting balance of concept and intention. TDC is currently and continuously pursuing original furniture and product designs, as well as offering a flexible range of design services. Their method of aesthetic and emotion can be applied in 2D and 3D, with print, objects, furniture, and spaces.

Page 198-199, 222-223

THEO HUMPHRIES (3EYES) AND PHILIP WORTHINGTON

Theo Humphries is the cofounder of 3eyes Design Consultancy where he collaborates internationally with creative people, concentrats on commercial and academic work, research, and events. He strives to provoke reactions in people; to promote questioning and reflection in others.

Philip Worthington founded Az2 Lab in 2002. His work spans graphic, product and interior design. He works creating interactive installation artworks as well as the development of collaborative networked applications. He has been exhibited around the globe at prominent galleries, festivals and museums and featured across numerous industry publications.

Page 156-157

TIM PARSONS

Tim Parsons was brought up in Wiltshire, England and studied industrial design at Royal College of Art in London from 1998 to 2000. Upon completion of his studies he began selling his own products to London retailers and undertook a research project for the Helen Hamlyn Research Centre at the Royal College of Art. In 2002 he moved to Manchester where he now combines teaching design at Manchester Metropolitan University with running a small studio. Mixing influences from craft and industrial design, his approach examines notions of familiarity, functionality and the quality of materials and processes, producing simple, durable objects.

Page 040

TOBIAS WONG

Born in Vancouver, Canada in 1974 but a resident of New York for nearly a decade, Tobias Wong belongs to a new generation of conceptualists who have purposefully blurred the boundary between art and design. His works are frequently disconcerting and amusing (sometimes in an off-colour manner) but always provocative and often beautiful and finely detailed. He has been recently awarded the 2006 Modernism/Brooklyn Museum Young Designer Award for Excellence in Design.

Photo: Dean Kaufman
Page 098-099

VITAMIN

In 2005 the brothers Chris and Andy Vernall decided to branch out from their design consultancy V2 Studios and set up Vitamin creating their own range of products which provides quirky solutions for 21st century living.

Vitamin's point of difference is in its philosophy to combine function and beauty. Vitamin looks at problems, people and products from a completely fresh perspective in order to produce exciting, pioneering and aesthetically pleasing solutions to enhance consumers' lives.

Page 009, 056-057

VLAEMSCH ( )

Vlaemsch ( ), a brand-new design label, has recently been established by Casimir, one of the leading Belgian designers of his generation. Vlaemsch ( ) stands for Flemish traditions humour, design, simplicity, sensible, the surreal, no-nonsense, quality & solidity and combines creativity with the highest level of quality: quality of design and quality of product.

Page 024-025, 050-051, 112, 119

VLIEGER & VANDAM

Vlieger & Vandam is a fashion accessories label based in Rotterdam, The Netherlands. Founded by Carolien Vlieger and Hein van Dam, they have worked together on various projects since 2000. Combining the skills of a graphic designer and a 3D designer, their work results in unique solutions.

'About three years ago we started making bags. We experienced that bags are perfect items to express your feelings, to make a statement. Vlieger & Vandam works with themes of everyday life. We get inspired by the six o'clock news and by exploring urban landscapes.' says Carolien Vlieger and Hein van Dam.

Photo: Marcel Kentin
Page 114-115

YAEL MER

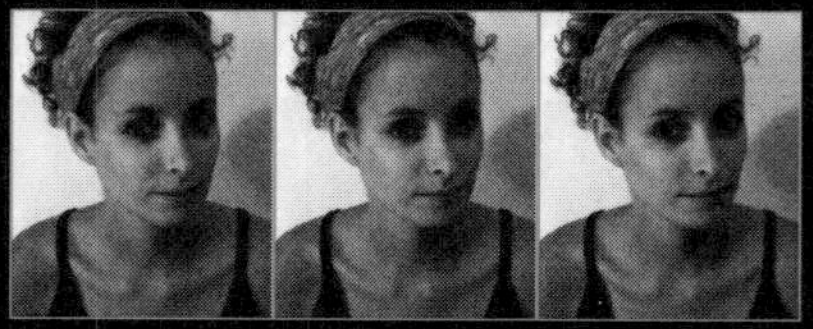

Yael Mer is a product designer graduated from the Royal College of Art, London.

Page 110-111, 208-209